A DANGEROUS WORLD

A DANGEROUS WORLD

Mark Raven, MD, MPH

HUTTMANN COMPANY

ANN ARBOR

Manufactured in the United States of America

1234567890

ISBN: 978-0-9617588-5-1
ISBN: 978-0-9617588-6-8

Library of Congress Catalog Card Number: 2009936709

Photo Credits

Spencer Platt/Getty Images News/Getty Images
Visual News/Getty Images
Time Life Pictures/Getty Images
Teh Eng Koon/AFP/Getty Images
Behrouz Mehri/AFP/Getty Images

Acknowledgement

The author wishes to thank Mary Walsh of Grubstreet for her editorial services and other valuable contributions to this book.

To My Family

Contents

INTRODUCTION

We have not had a world war for sixty years. We have not had a catastrophic worldwide epidemic since 1918. More recently, we have experienced regional disasters, such as Hurricane Katrina in our country and famines elsewhere in the world, but nothing on a national or worldwide scale.

Which disaster will come next? Will it be a natural disaster or man-made? Will there be a devastating influenza epidemic? Will we experience a total economic collapse accompanied by widespread civil disorder? Will there be a major jihad offensive by the radical fundamentalist Muslims against the West? Will there be a war between China or Russia and the United States? Will North Korea or Iran explode a nuclear weapon over the United States? Will Pakistan's nuclear arsenal be used to demolish our cities? Will there be a major conflict between the Globalists and the Nationalists? Whichever disaster comes first, the survivors will have other major battles to fight. There appears to be no end to potential conflicts.

I have written this book as a physician, trying to prevent or minimize human pain and suffering. We live in a dangerous world, and if you want your loved ones and yourself to survive, you will have to be prepared.

I have attempted to prognosticate potential man-made disasters based on current conditions. I have tried to do so without favoritism toward any political party, religious group, or ideology by presenting the positions of both sides of a potential conflict. The purpose of this book is not to take sides but to inform the reader of the various types of disasters that may occur at any time and to encourage everyone to prepare for these events so they may survive and suffer a minimum of hardship.

PART 1: A DANGEROUS WORLD

We live in an interesting but dangerous world. There are numerous threats to our safety, security, and way of life. For most of us, there is not much we can do but follow the course of events and do our best to provide safety and security for our families and ourselves.

The purpose of this book is to increase your awareness of the problems, give you the tools you need to assess your needs, and provide you with the information to protect yourself and your family to the greatest degree possible. The first step is to understand the risks. The second step is to assess your vulnerability. The third step is to determine what you can do within your means and situation to lower your risk of dying or getting injured when and if disaster strikes. The final step is to do it!

Preparedness is not a term reserved for fringe groups, religious entities, or militia groups. It is a term that applies to all citizens. Recall the recent public service announcements on television featuring presidents George H. W. Bush and William J. Clinton together urging the American people to prepare for any and all emergencies. They did not elaborate regarding what these emergencies might be, but this book does. Two former presidents of the United States would not go to the trouble to warn us if they were not worried about the future. The Department of Homeland Security maintains a Web site (www.ready.gov) encouraging and instructing citizens to be prepared. It is a useful reference covering material found in part two of this book. The Web site does not, however, go into detail as to the nature of the risks, and it does not have a scale upon which to rate your personal and family vulnerability. Those areas are covered in this book.

FEMA, the CDC, the FBI, and other federal and state government agencies are working diligently to prepare for disaster, whether natural or man-made. But individuals must take responsibility for their own safety. We cannot count on the government to take care of us in every situation (remember Katrina). Some disasters may be so devastating that there will not be enough resources to go around. Each person, each family, and each group of people must be prepared to take care of themselves. Self-reliance is essential to survival.

Most people live their lives in a confined space. We limit our concentration to those problems that are closest to us: problems with our families, our homes, our jobs. We worry about looking good, interpersonal relationships, paying bills, job performance, and job security. Most of us don't have much time left over to worry about broader subjects that may in the future greatly affect our lives—things like geopolitical changes, man-made and natural disasters, the depreciating dollar, radical militant religious sects bent on converting the entire world to their religion and eliminating nonbelievers. The list goes on and on. It is time to sit up and take notice.

Numerous serious threats to our safety, security, and even our survival are converging toward a perfect storm. Our economy has been gutted. Bank closures, civil disorder, and even revolution are potential outcomes. Islamic terrorists have silent cells in our country; they are waiting for a signal to strike again. Socialist countries south of our border are partnering with each other and with our longtime adversaries Russia and China. Iran and North Korea are developing nuclear weapons and guided missiles.

Is it just a matter of time? What will happen first? Will it be total economic collapse with destruction of the dollar, bank closures, and civil disorder? Will the radical Islamists strike again, this time on a much larger scale than 9/11? Will we have an internal conflict between the Globalists and the Nationalists? Will climate change increase the risk of tornadoes, hurricanes, or floods?

Will our public utilities be disabled? Will our food supply be interrupted? Will you be caught off guard? Will you be able to survive? You must understand the threats, how serious they are, and how likely one or more of them are to occur. You must prepare for your survival.

Disaster preparation varies somewhat on the type of disaster you anticipate. Will the disaster be of long or short duration? What amenities will be temporarily or permanently wiped out? Will you have to evacuate the area? These are all considerations you must take into account in your preparations.

The best plan is to prepare for the type of disaster that is most likely to occur in your geographical area. Do you live in an area likely to be hit by a hurricane? Or a tornado? How about a flood? What about a terrorist attack?

A basic understanding and awareness of the threats we face from man-made and natural disasters is essential if we are to prepare adequately for them.

Let's examine these threats in an orderly fashion. First, we will start with man-made disasters. Then we will look at the world's trouble spots. Finally, we will discuss global warming and natural disasters. Several types of disasters overlap these categories of natural versus man-made ones.

A. Precursors to Man-Made Disasters:

Overpopulation

Man-Made Disasters are those caused by human actions. There are four precursors of this latter category: overpopulation, economic conditions, ideological differences, and environmental pollution.

It took from the dawn of mankind until the early part of the nineteenth century for the population of the earth to reach one billion. At the end of the last ice age, about 10,000 BC, there were only four million people living on the entire planet. They lived mainly as hunters and gatherers. Two thousand years later, with the domestication of animals and plants, the population increased to five million. It took another eight thousand years, and the development of bronze and then iron, for the population to grow to 250 million. The Roman Empire was flourishing at that time.

Periodically decimated by major outbreaks of plague, smallpox, tuberculosis, leprosy, typhoid, syphilis, and other diseases, it took until the mid-1500s for the earth's human population to grow to 450 million. The plague (Black Death) killed 75 million people two hundred years earlier.

Smallpox vaccination began in 1796. Food canning was developed about 1810. Around 1825, the world's population finally grew to one billion.

The Irish potato famine occurred in 1846. The hypodermic syringe was invented in 1853. Dynamite was invented in 1866. The first automobile was built in 1893, and the first airplane was flown in 1903. By this time the population had grown to 1.6 billion.

From that point, population growth was gaining momentum like an accelerating freight train. With advances including the development of penicillin, a vaccine for polio, and the development of crop dusting with insecticides, it took only fifty more years for the world's population to reach three billion. The year was 1950—only sixty years ago.

The population explosion freight train was now running at full speed. By 1984, thirty-four years later, the earth's human population increased to five billion, and by 2000, it reached the astounding figure of six billion. The population had doubled in fifty years.

According to the current projections of the U.S. Census Bureau, the earth's population will reach 8 billion by the year 2025, and a staggering level of 9.3 billion by 2050. The majority of this growth is projected to occur in the lesser-developed countries of Africa, Asia, and South America.

Man, like other animals, has the remarkable capacity to develop from a single-celled embryo to a complex organism. Although man's intellectual abilities exceed those of other animals, most people never develop their intellects to their fullest extent. The brain must be trained through education and discipline to take full advantage of man's unique capabilities. Basic survival needs, including food, shelter, and reproduction, keep most people from attaining the maximal development of their intellect. Early man had to spend most of his time fulfilling these basic needs.

In our contemporary civilization, most people still have to spend much of their time meeting the basic needs of life. On the other hand, extensive specialization and division of labor have allowed some people to pursue education to its fullest extent. These relatively few people have led us to the brink of a supercivilization.

We live in the greatest technological age that man has ever known. We have conquered diseases that used to kill millions of people prematurely. We have made survival infinitely easier and allowed man leisure time for learning, enjoyment, and fulfillment of his desires and dreams.

We have sent man beyond our own planet to the moon, and we have explored even farther with our space probes. Data from the space programs indicate that the other planets in our solar system are inhospitable to life. These findings undermine previous conceptions regarding the planets, including thoughts of colonizing Venus and Mars. We can no longer entertain images of races of giants inhabiting Jupiter or supercivilizations existing elsewhere in our solar system. Perhaps, billions of years ago, other planets in our solar system may have been capable of supporting life, and perhaps, billions of years in the future, some other planets will have a life-supporting capability. But right now, we are here, alone, and there is no place else for us to go. If we destroy our planet's delicate balance of nature and its ability to sustain life, we shall also destroy ourselves.

The development of mankind is gravely threatened by one glaring fact: there are too many of us. Leonardo di Vinci once said that most men contribute nothing to the development of civilization. They only exist and in the process are detriments to society. They eat, produce waste, and when they die, they leave a decomposing corpse behind. The statement is still true today but on a much larger and more serious scale. Most people today consume vast amounts of our limited energy resources while contributing very little to the progress of civilization. In the process of using these energy resources, whether by driving automobiles or using modern machines or their products, they are in turn polluting the environment to a degree that threatens their own existence.

We are approaching a world population of seven billion people. They all want to enjoy the same standard of living that is prevalent in the United States. Oil reserves are declining, forests and farmlands are disappearing, and food stocks are at their lowest levels in years. There are not enough resources to support everyone at the level enjoyed by the United States.

The premise that we might some day pay the piper for overpopulation is certainly not a new one. We are indebted to two gentlemen for postulating this theory over one hundred years ago. They are Thomas Robert Malthus and Charles Darwin.

Malthus was an English economist and demographer who lived from 1766 to 1834. The Malthusian theory states that population tends to increase at a geometrical ratio, while the means of subsistence increase comparatively slowly at an arithmetical ratio, and that this leads to an inadequate supply of the goods supporting life. Historically, there have been three mechanisms for restoring balance whenever the human population exceeds the food supply: famine, disease, and war. In 1798, Malthus explained that without these periodic checks, the birth rate would so far exceed the death rate that the multiplication of mouths would nullify any increase in the production of food.

Malthus said that the only way to check overpopulation was sexual restraint. As far as economics was concerned, he was a pessimist and viewed poverty as man's inescapable lot. He pointed out that issuing relief funds or supplies to the poor encouraged them to reproduce all the more and thus add to the problem. Such an attempt at solution would merely postpone the calamity.

In nonindustrial societies, the Malthusian theory seems to have a potentially threatening validity. Industrial societies, however, have so far refuted the postulations of Malthus. National income has tended to outpace population growth, and the size of the family has become a matter of choice due to the prevalence of the various types of birth control.

It is interesting that it was Malthus who set Charles Darwin on the train of reasoning that led to the theory of natural selection. Darwin (1809–82) was an English naturalist who first established the theory of evolution in his monumental work, *The Origin of Species*. From 1831 to 1836, Darwin sailed in *HMS Beagle* as a naturalist for a surveying expedition. His observations of the relationships between animals separated by time and distance led him to reflect on the contemporary prevailing view of the fixity of species.

In October 1838, he read Malthus' *Essay on the Principles of Population.* His own observations had convinced him of the struggle for existence; upon reading the views of Malthus, it at once struck Darwin "that under these circumstances favorable variations would tend to be preserved, and unfavorable ones to be destroyed. The result of this would be the formation of a new species." Hence, the theory of natural selection, or survival of the fittest, was born.

In the nonhuman animal world the *fittest* individuals or groups are those whose physical and behavioral traits make them best able to survive in their environment. This is also true of humans, yet man is different. Man lives under social and cultural as well as biological rules. His technology, combined with his intelligence, allows him to survive under conditions that would kill him if he were an animal. Man's intelligence can manifest itself in many forms: the ability to comprehend complex situations and to solve problems; military acumen; scientific, mathematical, and engineering ingenuity; business acumen, and so on. Intelligence combined with the basic instinct of aggressiveness is an especially ideal combination for fitness and survival in the human situation.

The economists and demographers of today say that Malthus was short-sighted because he could not anticipate the development of modern agricultural techniques like disease-resistant plants, fungicides, and insecticides, all of which have increased food production markedly. He also did not anticipate the advances in science, engineering, and medicine, which, over the years, have brought remarkable increases in our standard of living and allowed us to live longer and healthier lives.

Ironically, the industrialization that has given us a high standard of living has resulted in deforestation and pollution of land air, and water. Pollution has resulted in contamination of major food supplies, disappearance of animal and plant species, and major illnesses among workers and consumers exposed to the toxic by-products of industries. The United States has led the way in reducing pollutants in the workplace and protecting the environment though at the cost of losing jobs to other countries where antipollution laws are less strict or nonexistent.

But we are again approaching the end of our rope. We are running out of resources. In short, there are too many people and too many machines, all consuming too many goods. We have two choices. We can cut down on the number of people, reduce environmental destruction and pollution to a viable level, and progress toward a supercivilization, or we can continue as we are and eventually destroy ourselves. The choice is ours.

Will we be able to develop new resources sufficient to support eight billion people? Or will we have a massive population cleansing by war, famine, or disease and reduce the demand on the earth's resources to sustainable levels?

The bottom line here is that the conflicts described later in this chapter are offshoots of the root problem—overpopulation. The pressures of overpopulation on the earth's ability to sustain it are approaching the breaking point. The usual equalizers of overpopulation may soon recur in force. If it is war with weapons of mass destruction, the reduction in population may be swift but furious, and massive in scope. Only those who are prepared will have a reasonable chance to survive.

Economic Decline

From the economic viewpoint, we are rapidly plunging into a dark age. This is true not just in the United States but all over the world. The causes of the deteriorating economy are complex and beyond the scope of this book. It may be noted, however, that just as government policies can cause famines, government manipulation can cause economic depressions and create other crises that enable those governments to change laws and manipulate its citizens. Suffice it to say that recession is here, and depression is just around the corner. You need to prepare for it if you are to survive.

Distrust is everywhere. The same politicians and executives of banks, brokerage houses, and insurance companies who caused this economic crisis are in cahoots with each other and are still in power. They are responsible for solving the crisis and are trying to cure the problem with more of the same destructive tactics, further bankrupting the country by giving away trillions of newly created dollars to the very institutions that created the worthless derivatives in the first place. Instead of being punished for their economic misdeeds, these executives receive bonuses and bailout money.

Many pension funds have discovered they have worthless derivatives among their investments. Caught in the stock market crash, they may not be able to meet their obligations to their retirees. Hedge funds have lost billions of their investors' money. Municipalities have lost the tax money they have invested and now cannot meet their financial obligations. Corporations have also been caught in this financial web.

Because of globalization, millions of jobs have been exported to countries where salaries of workers are a fraction of those in the United States and antipollution laws are sparse or absent. As a result, unemployment in the United States is skyrocketing, and workers have no money to spend. Retail sales are in a tailspin.

As of April, 2009, the real unemployment rate is not 8.5 percent as stated by government statistics. That figure does not count the part-time employees who don't make enough to live on. Neither does it count those who have given up hope of finding a job and have stopped seeking employment altogether. The real unemployment is around 15 percent. And the rate continues to rise.

When people cannot meet their daily living expenses, they use their credit cards and go into debt. They stop paying their mortgages and car payments and eventually lose their homes and cars. Tent cities spring up around the country. Emotions run high and protests get underway.

In some larger cities, the high-school graduation rate has dropped below 50 percent. That is a staggering and alarming figure. These young people are generally unemployable. Angry and desperate, many of them join gangs and take up a life of crime and drug addiction.

The dollar is being devalued by credit creation and excessive money printing. In the not-too-distant future, deflation will be replaced by hyperinflation, and the remaining life savings of most citizens will lose purchasing power.

A bank holiday was declared and banks were closed for a period of time during the Great Depression. The government called in gold, which was officially used to back the U.S. dollar at that time (at $20 per ounce), and later reissued it at $35 per ounce. There is no gold backing the U.S. dollar today. It is only worth the paper upon which it is printed and the confidence people have in its worth.

With the present crisis, a bank holiday may be called at any time because most banks do not have the assets to meet their obligations to the depositors. If everyone tried to withdraw their savings at once, most banks would not have enough money to pay them. In such a situation, a bank holiday would result in the freezing of your accounts. Credit cards would become nonfunctional, and assuming your guaranteed deposits were eventually paid back, the purchasing power of those deposits would be greatly reduced because by the time you finally received the money, the dollar would be devalued considerably or even replaced with a new currency.

This current economic crisis may eventually result in civil disorder. Protests are already occurring. Riots, burning, and looting may follow. Public utilities may become nonfunctional. Under these conditions, civil behavior disappears and barbarism prevails. Economic collapse is a real threat and may occur soon. The need to be prepared for it is urgent.

Will you be able to protect yourself and your family if civil disorder commences because of the deteriorating economy?

Ideological Differences

Differences in religious and political ideologies are frequent causes of terrorism and warfare. The crusades of the Middle Ages were caused by religious differences between Christians and Muslims, and the battle between these religions continues today. Terrorist attacks by Muslim extremists against Jews occur frequently, and the possibility of all-out warfare between Muslims and Jews is particularly high. Political differences between Communism and Capitalism resulted in the Korean and Vietnamese wars, and although the differences are subdued for the present, they could flare up again at a moment's notice. World War II was caused by both political and religious ideological differences.

There are numerous ideological conflicts throughout the world today. The potential for both regional conflicts and worldwide warfare is as great as in any time in history.

Environmental Pollution

Greenhouse gases have received the lion's share of publicity in the press and in the speeches of politicians because this segment of the environmental pollution problem has become a political football game. Politicians see global warming as a way to drastically raise taxes and at the same time increase their power base. Many reputable scientists dispute the effect of these gases on global warming and believe the climate changes we are observing are due to causes other than carbon dioxide and similar gases.

While concentrating on global warming and greenhouse gases, politicians and the press have not paid much attention to another growing problem: the thousands of toxic chemicals spewed into the environment by industry, mining, and agriculture. These chemicals— asbestos, lead, mercury, polycyclic hydrocarbons, herbicides, and pesticides, just to name a few—have caused diseases of epidemic proportions and have contaminated a significant proportion of our food supply.

Since the 1960s, the United States has been a leader in cleaning up these chemicals both in the environment and in the workplace. But the cost of the environmental control systems has raised the cost of production. In response, manufacturers have exported millions of jobs overseas to countries where environmental pollution and worker exposure go on unabated. However, this does not reduce the overall effect on worldwide pollution. The pollution, including greenhouse gases, is only transferred to another area. Environmental diseases are increasing at a rapid pace in the developing countries.

Outsourcing is the equivalent of condoning sweatshops. We now consume products from China, Mexico, and other countries at the expense of their workers and citizens. The overall effect on worldwide pollution is unchanged

B. Man-Made Disasters:

Man-made disasters include wars, terrorism, nuclear-chemical-biological attacks, nuclear power plant emergencies, environmental contamination, and electromagnetic pulse (EMP) warfare. Several disasters overlap the natural and man-made categories, including famine (by government or economic policies), floods (poorly-constructed dams or faulty planning) and wildfires (arson or out-of-control burns). These last three categories will be discussed in the section on natural disasters. Terrorism will be discussed in the section on world trouble spots.

Conventional Warfare

Wars have decimated populations throughout history. Fifty-five million people died as a result of World War II. Eleven million died during World War I. With the exception of the atomic bombs dropped on Hiroshima and Nagasaki, these wars were fought with conventional weapons. The U.S. Census Bureau estimates the world population at that time to have been 2.3 billion people. According to these figures, 2.4 percent of the world's population died during that war.

The total number of Americans who have died in wars is estimated to be 1,314,000. The Civil War (1861–65) claimed the most (623,026), with World War II (1941–45) the second highest (407,316). World War I (1917–18) claimed 116,708, while the Vietnamese War (1964–1973) recorded 58,169 deaths; in the Korean War (1950–53) the death toll reached 36,914. These numbers do not include those who were injured.

The next major war may not be fought with conventional weapons. With weapons of mass destruction, millions could be killed or injured within hours. Cities could be totally destroyed. The environments around these cities could be rendered uninhabitable for generations due to lingering radiation or chemical contamination. Food and water supplies could be rendered unfit for human consumption, also for generations.

Rogue states, terrorist groups, and religious fanatics, all soon to have or currently having access to nuclear, chemical, and biological weapons, pose the greatest threat imaginable to modern civilization. These threats are the hardest to combat in terms of survival. They require the greatest degree of preparation in terms of expense and training.

NBC Warfare

A nuclear, biological, or chemical attack, whether by terrorists or in all-out warfare, may have devastating consequences. The number of dead and injured could be very high, reaching up to millions. Responding personnel will be at risk of injury, contamination, and death. The psychological effect on the population will be overwhelming, causing fear and panic, as well as posttraumatic stress syndrome.

The threat is real. Pakistan has a nuclear arsenal, and Iran is developing nuclear capability. North Korea has already conducted nuclear test explosions. These countries have delivery systems in place or in the process of development. Russia has had nuclear weapons for many years. Smuggling of Russian nuclear weapons has been suspected. Russia also has had an extensive biological weapons program for many years, and has weaponized numerous infectious agents. Other countries which support terrorism are suspected of having similar programs. Among those agents known or suspected of being weaponized are anthrax, smallpox, ebola and other viruses, as well as toxins such as botulism and ricin.

Chemical agents, including nerve gases, cyanide, chlorine, phosgene, and mustard gas have been used in warfare for many years. These agents are easier to manufacture and deliver than nuclear or biological weapons.

Nuclear Warfare:

Nuclear warfare is the most devastating of the weapons of mass destruction. Biological and chemical weapons kill and injure people but usually leave the infrastructure of civilization intact. Survivors are able to carry on.

Nuclear warfare not only kills and maims large numbers of people, it destroys or contaminates infrastructure, making areas uninhabitable for years. It destroys cities, farmland, and forests. Individuals surviving the initial blast may develop ARS (Acute

Radiation Syndrome), and there is long-term risk of developing cancer due to excessive exposure to radiation.

From the limited dirty bomb, an explosive device laced with radioactive material, to a massive hydrogen bomb, residual radioactive material deposited from an explosion can render an area uninhabitable for years. Apart from suitcase atomic bombs, dirty bombs are weapons most likely to be used by terrorists. Larger bombs are more sophisticated and are more likely to be used in wars between countries.

On August 6, 1945, the Enola Gay dropped the first atomic bomb on Hiroshima. A small bomb by current standards, it destroyed the city and initially killed 100,000 residents; another 40,000 died of burn complications, radiation sickness, and other related causes by the end of the year. The blast completely leveled 48,000 buildings, and only 6,000 of the original 76,000 buildings were left intact.

If you are near an explosion involving nuclear material and survive the initial blast, radiation exposure will be your main concern. You must limit your radiation exposure as best you can. The three factors you can use to protect yourself are shielding, distance, and time. Shielding between you and the source of the radiation will limit your exposure because the lead or other shield material will absorb part of the radiation. Distance is an important factor because your exposure is lessened considerably the farther you are from the source. It is also important to limit the time of exposure. The quicker you can evacuate the area, the better.

Biological Warfare:

Biological agents are probably the oldest of the nuclear/biological/chemical weapons of mass destruction. They have been used for over 2,500 years. They are more calamitous than chemical weapons.

Inhalation of the offending organism is the primary route of transmission. Biological weapons are usually dispersed as an aerosol. They may be disseminated from a point source such as industrial sprayers, from moving sources such as airplanes or boats, or from military weapons such as bombs or missile warheads. Humans are not the only potential targets of biological warfare. These agents may also be used to destroy crops and farm animals.

Between fifteen and twenty countries are thought to possess biological weapons, including Iran and North Korea. Russia has had an extensive and sophisticated biological weapons program for many years.

The ideal biological weapon can be delivered as an aerosol and maintains its viability and infectivity once dispersed in the environment. It has a high disease to infection ratio and has a vaccine or other method available to protect the attacker.

The purpose of a biological weapon is causing widespread illness and death. It causes large numbers of casualties and severe psychological stress on the population; it creates extensive need for medical services, overwhelming doctors and hospitals. It also creates, as an added burden, the need for individual and collective quarantine and protection of medical personnel.

In an article published in the prestigious journal, *Science*, Dr. D.A. Henderson informs us that of the thousands of biological agents that could be used as weapons, only a few make near-perfect weapons with high infectivity, ease of production, ability for aerosolization, high death/disability rates, human-to-human spread, and other factors. Experts have determined that the two biological weapons most likely to be used by bioterrorists are smallpox (a virus) and anthrax (a bacterium). Other biological agents under consideration include bacterial agents like bubonic plague and tularemia, viral agents like Ebola and VEE, and the toxins botulism, ricin, and SEB (Staphalococcal Enterotoxin B).

Smallpox:

Smallpox vaccine was developed by Edward Jenner in 1796. Smallpox is again a serious threat because vaccinations against the disease stopped more than twenty years ago, and very few people still have immunity. The virus can survive for twenty-four hours or more in an aerosol form, and it is highly infectious even in small amounts. There is no treatment. Since smallpox is a virus, antibiotics are of no use. A second wave of cases, occurring about 14 days after the first wave, is almost inevitable. This second wave will decimate the medical population of doctors, nurses, and other medical personnel who will come in contact with the first wave.

If adversaries of the United States are considering using smallpox as a weapon against its people, it will not be the first occurrence of its kind. Smallpox was widespread at the beginning of the American Revolution. There are indications that the British had purposefully sent infected blankets into American Indian camps during Pontiac's War (1763). Epidemics occurred among the tribes, killing more than half the population. American soldiers besieging Boston and Quebec in 1775 were convinced of the same sort of British treachery.

In 1777, Major Robert Donkin, a British officer during the American Revolution, published a book called *Military Collections and Remarks* (published by Hugh Gaine, New York City). On page 190, there is a footnote, which is deleted in many copies because of its inflammatory nature but is still present in a copy obtained by the Clements Library at the University of Michigan. The footnote reads:

"Dip arrows in matter of small pox, and twang them at the American rebels, in order to inoculate them; This would sooner disband these stubborn, ignorant, enthusiastic savages, than any other compulsive measures. Such is their dread and fear of that disorder!"

Smallpox is a deadly disease. The usual mortality rate for a person without prior immunization stands at 30 percent. In some epidemics, however, it has been as high as 50 percent. Smallpox is thought to have killed more human beings than any other infectious disease. It is known to have occurred among the early Egyptians. It killed at least three hundred million people during the twentieth century.

Smallpox has an incubation period of ten days. The first symptoms are fever, backache, and vomiting. Small red spots then appear all over the body, though more densely on the face and extremities, and the patient gets a characteristic worried look. The spots then turn into painful blisters (pustules) with a dimple in the center. The pustules become larger and fill with virus-containing fluid. The pustules then become hard, about the size of peas, giving the skin the appearance of a cobblestone street. If the patient lives, the pustules leave scars for the remainder of the patient's life.

In 1980, the Soviet government began a successful program to create an even more virulent strain of the smallpox virus and to weaponize it so it could be released on an enemy using bombs and missiles. There is concern that the weaponized virus has been passed on to other countries.

Smallpox spreads from person to person by droplets or aerosol expelled from the nose or mouth of infected persons. Contaminated bed linens or clothes can also spread the virus. Smallpox patients are most infectious from the onset of the rash through the next seven to ten days.

The above-mentioned Dr. Henderson is the man credited with ridding the world of smallpox. Due to the efforts of a team led by him, smallpox was eradicated from the world in 1979. It is a fascinating story:

On February 15, 1972, a thirty-eight-year-old Muslim clergyman returned to his hometown in Kosovo from a pilgrimage to Mecca. The next morning he awoke feeling achy. He shivered for a day or two, then developed a red rash. He had been vaccinated for smallpox two months earlier. Yugoslavia was well vaccinated; the last case of smallpox had occurred in 1930. The clergyman's family and friends came to visit him. Unknowingly, he was breathing smallpox virus into the air, attached to droplets of saliva. These were inhaled by the visitors.

Several weeks later on March 3, 1972, one of the visitors, a thirty-year-old male schoolteacher developed fever. Two days later, he developed dark spots on his skin, which turned into blackened, mottled splashes; the local doctors did not recognize the condition as hemorrhagic smallpox. He developed bleeding in the sclera of each eye. On March 10, 1972, he suffered massive hemorrhages in the intestines and expelled quarts of blood along with the sloughed intestinal and rectal mucosa; he died from exsanguination.

The doctors, nurses, and some of the patients at the hospital who cared for him became infected. Meanwhile, the clergyman infected twenty-seven people at the hospital alone, and thirty-eight people in all. Eight of them died.

A World Health Organization team, led by Dr. Henderson, rushed in to contain the epidemic. The Yugoslavian Army was mobilized, villages were closed, roadblocks were erected, and ten thousand people were quarantined by the Yugoslav military. The country's borders were closed. A massive campaign was instituted to revaccinate every person in Yugoslavia. Eighteen million doses of vaccine were given in ten days. The epidemic was contained, and smallpox was finally eradicated from the human population.

The release of a highly virulent weapon-grade smallpox virus into the population would constitute a global emergency. The disease would spread in logarithmic fashion, with each diseased person infecting an average of twenty people. As a result, within ten to fourteen days of the first occurrence, the infection would spread from 1 to 20, then to 400, then to 8,000, then to 160,000, and so on. Millions, perhaps billions, would die from the disease.

A genetically manipulated, highly virulent, highly transmissible strain of smallpox is the ideal weapon for a country or a religious or politically inspired group to use to eliminate their adversaries, nonbelievers, or opponents. After vaccinating all their followers and then releasing the virus into the general population, they would be the only major surviving group.

The main defense against contracting the disease, if it is released into the population, is quarantine—complete isolation from that population until the epidemic has run its course. There may not be time to develop a vaccine against the disease.

Anthrax:

Anthrax is an infectious disease that occurs naturally in cows, sheep, goats, camels, and other animals. It can also occur in humans by handling products from infected animals, inhaling anthrax spores from contaminated animal products, or by eating undercooked meat from infected animals. Anthrax has also been referred to as "woolsorters' disease." It is caused by the spore-forming bacterium, *Bacillus anthracis*. Anthrax spores can live in the soil for many years.

The disease can occur in three forms: infection of the skin, the lungs, or the intestines. Symptoms usually occur within seven days. However, with inhalation, anthrax symptoms can take up to forty-two days to appear. Antibiotics are used to treat all three types of anthrax. Early identification and treatment are important. Before the symptoms appear, Anthrax can be prevented after exposure with antibiotics plus anthrax vaccine. After the symptoms occur, the treatment is a sixty-day course of antibiotics. Treatment is not always effective, and patients may die regardless of the medication.

With skin infection, the spore becomes a cut or abrasion; it is usually contracted in handling of contaminated wool, hides, hair, or leather of infected animals. The infection starts as a small sore that looks like an insect bite, but it soon becomes blisterlike and then a painless ulcer about an inch in diameter with a characteristic black center. About 20 percent of patients with cutaneous anthrax die if untreated.

With lung infection, the initial symptoms are sore throat, mild fever, and muscle aches. After several days, the symptoms progress to severe breathing problems and shock. Inhaled anthrax is usually fatal, even with aggressive antibiotic and supportive therapy. The inhalation route is the most likely route that would be used by terrorists. The spores can be disseminated from a spraying device either on the ground or from an airplane or helicopter.

Intestinal infection usually occurs due to consumption of contaminated meat. It causes acute inflammation of the intestinal tract. Initial symptoms include nausea, loss of appetite, vomiting, and fever; these are followed by abdominal pain, vomiting of blood, and severe and sometimes bloody diarrhea. Patients may complain of sore throat and difficulty in swallowing. Swelling of the neck and regional lymph glands may occur. An average of 25 percent to 60 percent of patients with intestinal anthrax die.

Anthrax was used as a biological weapon in the United States in 2001. Anthrax spores were deliberately spread through the postal system by a scientist who worked at the government's biodefense labs. He sent letters containing anthrax spores to various governmental employees and others. Twenty-two cases of anthrax infection resulted from the mailings, eleven of which were cutaneous anthrax and eleven inhalation anthrax. Five of the inhalation anthrax cases died from the disease.

Government agencies are working diligently to prepare not just for anthrax but for all kinds of possible bioweapon attacks. Plans and procedures are in place to respond to an attack. Emergency response teams have been trained and equipped to help state and local governments control infection, obtain samples, and perform laboratory tests. These agencies have developed programs to educate health care providers, the media, and the general public regarding potential attacks. The Centers for Disease Control (CDC) works closely with health departments, veterinarians, and laboratories in surveillance efforts; it also works with hospitals, laboratories, and emergency response teams to make sure they are equipped to handle bioweapon attacks.

Chemical Warfare:

Primitive forms of chemical warfare were used in ancient times to poison water supplies or to expose an enemy to noxious smoke. Chinese writings contain hundreds of recipes for the production of poisonous or irritating smokes for use in war. The poisoning of wells during the time of the Roman Empire has been documented.

During World War I, Germany used chlorine gas against the Allied troops in Belgium in 1915. Five thousand troops died, and ten thousand became sick from the gas. Other chemical warfare attacks during that war included the use of phosgene and mustard gas.

Classes of chemical warfare agents include nerve agents, (Tabun, Sarin, Soman, and VX), blood agents (hydrogen cyanide and cyanogen chloride), choking agent and lung irritants (phosgene and chlorine gases), vesicants or blistering agents (mustard gas and Lewisite), as well as miscellaneous gases and sprays such as ammonia, Mace, pepper spray, and others. Nerve agents are highly lethal. They cause convulsions, paralysis, and death by respiratory failure. They are similar to the organophosphate pesticides. Blood agents block the uptake of oxygen into the cells. They were used in extermination camps during World War II. They are also thought to have been used in the Middle East in the 1980s by both Syria and Iraq. Choking gases and lung irritants cause formation of water in the lungs, which interferes with the oxygen transport from the lungs to the bloodstream and ultimately leads to death. Vesicants or blistering agents cause severe blisters on the skin. They also affect the eyes and lungs. Death may occur from toxic shock within twenty-four hours of massive exposure.

The level of knowledge required for manufacturing chemical weapons is much less than that required for making biological ones.

Many chemical warfare agents can be made by a chemistry graduate student. Most of the materials and equipment are relatively inexpensive and can be commercially purchased without notice or alarm. These agents are ideal weapons for terrorists. Two recent cases of chemical warfare or terrorism are described below.

In 1985, a religious cult in the Ozark Mountains of Arkansas obtained a thirty-gallon drum of potassium cyanide. Their plan was to poison the water supplies of New York, Chicago, and Washington DC. They believed they were on a divine mission and that these attacks, which were intended to punish heathens and unrepentant sinners, would hasten the return of the Messiah. However, the cult members lacked the scientific and medical background to fully execute an effective attack. The attempt did not materialize because prior to the planned attack, federal agents discovered the cyanide during a raid on the group's compound.

In 1995, members of Aum Shinrikyo, a quasi-Buddist religious cult in Japan, released the nerve agent Sarin in the Tokyo subway system. The attack killed twelve people and injured more than a thousand. Several cult members, trained university scientists, produced the Sarin in a laboratory at its main compound at the base of Mount Fuji. They had planned to carry out several attacks with chemical and biological weapons in Tokyo and other Japanese cities. The attack would have been more deadly, but the perpetrators did not vaporize the Sarin. They carried plastic bags of the agent in liquid form onto subway cars and then punctured the bags, allowing the liquid to flow across the floors. The liquid evaporated slowly and only killed the passengers close to the release sites. Had they used a vaporizer, the death toll would have been much higher.

The same cult had used Sarin in a less publicized attack in Matsumoto in 1994. They attempted to kill three judges who were about to rule against them in a lawsuit. The attack killed seven people and injured 144 others.

Chemical agents can be very effective as killing and disabling weapons both in the hands of terrorists and in all-out warfare. They are limited to localized or regional use when used as aerosolized agents. They can be used more widely as poisons in food or water supplies.

Nuclear Plant Disasters

The two most well-known accidents at nuclear power plants are the ones that occurred at Three Mile Island and Chernobyl. Of the two, Chernobyl was disastrous, while Three Mile Island was comparatively less so.

The Chernobyl disaster occurred in April, 1986. Located in the Ukraine and built with Soviet design, it was manned by poorly trained technicians. Safety features were lacking. There was a steam explosion and fire, which released part of the radioactive core into the atmosphere. The wind carried the radioactive material over the land, exposing the surrounding population to significant amounts of radiation. Large areas of Belarus, Ukraine, and Russia were contaminated with radiation in varying degrees. Lesser amounts were detected in Scandinavia and parts of Europe. Several hundred thousand residents of the immediate vicinity were evacuated and relocated to avoid lingering radiation. Twenty-eight people died from radiation and thermal burns in the ensuing four months. A total of fifty-six fatalities were reported up to the year 2004. By the year 2000, four thousand cases of thyroid cancer had been diagnosed in children from the exposed areas.

In addition to the potential physical health hazards of radiation exposure, the psychological impact of exposure can be widespread and devastating. Panic and mass hysteria are significant factors in dealing with an accident of this type. Post-traumatic stress syndrome can be present in a significant portion of the population. Lingering questions regarding cancer hazards in the exposed populations last for years.

Possibility of disasters must be considered in any area where a nuclear power plant exists. Even with the most sophisticated safety control systems, there is always potential for an accident at any time. Anyone living in the area should be prepared for immediate evacuation. Proper education of the population regarding the risks and the need for evacuation is essential for ensuring orderly withdrawal from the area.

Electromagnetic Pulse Attack

An especially worrisome scenario is the use of a nuclear weapon detonated in the upper atmosphere thirty to two hundred miles above the earth. No radiation or blast effect would reach the ground, but the intense radiation from the fireball would create a powerful electromagnetic pulse (EMP), which would damage integrated circuits and transistors in electronic devices over a wide area on the ground below the blast. There would be a tremendous power surge of thousands of volts through power lines, damaging any electrical devices connected to the power grid.

Computers would be fried. Automobiles and trucks using electronic ignition systems would be disabled, as would hospital-monitoring devices (vital sign-monitoring devices, ventilators, etc). Water and sewage treatment plants would not function.

Using a missile and one nuclear bomb, a rogue nuclear state such as Iran, Pakistan, or North Korea could send the United States back to the preindustrial era where there was no electricity. Airplanes would not fly, and defense systems would be disabled.

C. World Trouble Spots:

Radical Islam:

We are witnessing the resumption of a monumental battle between two cultures. Christians and Muslims have been fighting for centuries, and the current conflict is thought of by many as just a continuation of the crusades after a lull of several hundred years. A review of the history of the crusades exposes man's inhumanity to man. Barbaric acts—tortures of many varieties, including burnings at the stake, crucifixions, beheadings—all the result of religious disagreements, have terrorized populations for millennia.

This battle between Western civilization and a radical dictatorial theocracy has resumed and will only escalate as months pass. The storm clouds are brewing. The United States has already been viciously attacked (the World Trade Center attack of 9/11), and the perpetrators are threatening even more devastating and widespread attacks on this country and the rest of the world.

Fundamentalist Muslims (as opposed to moderate Muslims) pose a serious threat not only to the United States, but to all non-Islamic nations. Their ultimate goal is the conversion of the entire world to Islam and rule by governments based on Sharia law as propounded by the Koran. Their aims are rooted in history. An in-depth study of Islam will provide a better understanding of their mission.

Islamic fundamentalists have an intense hatred of the United States. Our way of life violates Islamic law and principles. They have an even more intense hatred of Israel and of all Jews. Their intent is to destroy both countries and all its infidel citizens. They believe in the use of violence to achieve their goal of a worldwide Islamic empire.

There are a number of terrorist groups that have established well-organized cells in the United States. Terrorism experts have estimated the number of Islamic fundamentalists in the United States to be at least one hundred thousand. Many of their members blend in with Western culture and become a part of communities and so are not suspected of being terrorists.

To learn more about Islamic fundamentalists and their goals, I suggest reading any of several noted authors on the subject, including

Yossef Bodansky (*Bin Laden: The Man Who Declared War on America*, Forum Prima, 1999), Walid Phares (*Future Jihad*, Palgrave MacMillan, 2005), and Steve Emerson (*Jihad Incorporated*, Prometheus, 2006).

Al-Qaeda:

Established about 1988 by Osama bin Laden, Al-Qaeda is an international terrorist network. In its early years, it was involved in the Afghan resistance against the Soviet Union. It has since become an international terrorist network with the goal of establishing an Islamic Caliphate not only throughout the Middle East but throughout the entire world. In 1998, Al-Qaeda issued a statement that it was the duty of all Muslims to kill U.S. citizens and their allies everywhere. In 2001, prior to the World Trade Center attack, Al-Qaeda merged with the Egyptian Islamic Jihad organization led by Ayman al-Zawahiri.

Al-Qaeda is responsible for assassinations, bombings, hijacking, kidnapping, and suicide-bomber attacks. Reports indicate that the group is attempting to obtain weapons of mass destruction.

The Muslim Brotherhood:

The Muslim Brotherhood is an Islamic religious and political organization dedicated to the establishment of a world based on Islamic principles. The organization was founded in Egypt in 1928. It began as a radical underground force in Egypt and other Sunni countries, but has spread worldwide. The group promotes strict moral discipline and opposes Western influence, often by violence.

For the last fifty years or so, Muslim Brotherhood has moved into Europe and the United States and has established a wide and well-organized network of mosques, charities, Islamic organizations, and training camps, with the ultimate goal of converting the entire world into a united Islamic state. Through these entities, they raise funds for terrorist activities and recruit new members.

Globalists versus Nationalists:

Differences in political ideology are polarizing the citizens of the United States. The differences have been present for many years, but until recently, the checks and balances system of government has kept one side's agendas from overwhelming the other. Now the balance of power is one-sided, and the dominant group is rapidly promoting their agendas; the opposition is almost powerless and is unable to stop them. Many of the dominant group's new policies, regulations, and laws are alienating the citizens of the opposing group. The two groups are not what you might first suspect. They are not Democrats versus Republicans.

The members of the dominant group call themselves "Progressives." Sometimes they are referred to as "Statists." Both terms are vague and misleading. Progressives believe that it is the government's responsibility to solve social problems and establish fairness in the distribution of wealth not only within the United States, but throughout the world. They cross party lines, some being Democrats, others Republicans.

A broader, more descriptive term is "Globalists." Many Globalists believe in a socialist one-world order. They envision a central government for the entire planet, usurping and replacing the power of all other governments, including the United States. In their view, nationalism is just a step in the evolution of civilization. They believe a one-world government is the wave of the future. They reject Social Darwinism and believe in an all-powerful government, socialism, and redistribution of wealth.

Over many years, the Globalists have infiltrated positions of power both in government and in the financial arena, and there has been intermingling between the two entities. Globalism is not a new phenomenon. The idea has been promoted for over a hundred years. Woodrow Wilson (founder of the Federal Reserve and the League of Nations) and Theodore Roosevelt were Progressives. Currently, the Globalists control or strongly manipulate all branches of government including the House of Representatives, the Senate, and the Presidency. They are very influential in the supreme court. They have the majority to change or manipulate laws to their favor. It has been a slow process, but they have finally become powerful enough to push forward their agenda at a furious pace and cause great consternation among the Nationalists.

Globalists now control most of the media, especially newspapers and television. They manipulate the news, so that the public mainly hears stories favorable to their agenda. Stories favorable to the opposition are frequently suppressed. Talk radio survives as a dominant voice for the Nationalists, and the Globalists are working hard to pass a fairness doctrine to suppress it.

For many years, the Globalists have also controlled the educational system from kindergarten to colleges and universities. Courses favorable to their agenda have been encouraged, while courses favorable to the Nationalists, including civics, have been suppressed. Many Nationalists have homeschooled their children to prevent them from being influenced by Globalist teachers. Teachers with a Nationalist bent have been ridiculed or ostracized by the majority.

Since the late 1990s, the Progressives (Globalists) have encouraged lower mortgage standards and urged banks to give loans to people with reduced credit ratings in order to promote their social agenda of redistribution of wealth and increasing the standard of living for everyone. This was the beginning of the current economic crisis. Whether they were too short-sighted to anticipate the economic catastrophe resulting from their actions, or whether it was part of a broader agenda to cause the catastrophe and allow them to later gain more control—"A good crisis should never go to waste"—is a matter of debate among the Nationalists.

Globalists have reached such power and influence in government that in May, 2009, the government-sponsored MIAC (Missouri Information Analysis Center) issued a report entitled, "The Modern Militia Movement," which stated that anyone protesting illegal immigration, supporting anti-abortionists, or even supporting Congressman Ron Paul (a staunch Nationalist) might very well be a member of a domestic paramilitary group and therefore a potential domestic terrorist.

The Nationalists, on the other hand, believe in the sovereignty of nations. They support the original Constitution of the United States and the Bill of Rights and believe these documents created a nation that has surpassed the wildest dreams of their authors. They promote self-reliance and the right of each person to reap the rewards of his own labor. They believe in individual freedom and rights. They believe that people have a right to bear arms and to protect themselves from tyranny, be it from foreigners or from their own government.

They are frustrated with the loss of jobs to globalization. They are becoming increasingly angry about the deteriorating economy, increasing taxes, and the lowering of the standard of living. They blame the government for causing these problems. Nationalists believe the Globalists have purposefully gutted the economy with the intent of creating a crisis that will enable them to pass laws favoring socialism and globalism.

The conflict between the Globalists and Nationalists is based not only on political, but religious ideology as well. A significant percentage of Nationalists have strong religious convictions and feel that the Globalists are against all organized religions.

These tensions may escalate and reach boiling point. Unless the two groups work out a mutually agreeable compromise, there is every possibility of civil disorder and even more serious consequences. A possible tipping point might be if the Globalists try to rescind individuals' rights to bear arms. Nationalists believe that they will then be at the mercy of their government or any other power that wants to control them. They are determined not to let this happen.

These ideological differences may cause devastating consequences for the country and may open the door for other adversaries of the United States to enter the fray for their own gain.

China and Russia:

Both China and Russia have long been powerful adversaries of the United States. There is now a delicate balance between mutual co-existence and a resumption of cold war tensions. China is building a formidable military system and has made threatening statements against the United States. Russia is helping Iran develop nuclear facilities.

During April, 2009, a national counterintelligence executive reported that Chinese cyber spies have penetrated deep into the U.S. defense network, the banking sector, and the electric grid. Counterfeit Chinese routers and chips may have even been incorporated into U.S. military aircraft. Bottom line: with specially programmed counterfeit chips, an adversary could have the ability to disable a system any time it chose.

Adversarial countries are banding together against us because they hate our way of life and they seek to destroy us. Even South American countries that used to be our friends are forming alliances with China and Russia.

Venezuela:

Venezuela has a population of about 27 million people and is about twice the size of California. The country's main natural resource is oil. In 2007, the Venezuelan president Hugo Chavez nationalized firms in the petroleum, communications, and electricity sectors. In 2008, he continued his efforts to increase the government's control of the economy by nationalizing firms in the cement and steel sectors. Chavez is the hub of anti-American sentiment in Latin America. He has an intense hatred of the United States and is forming alliances with other countries hostile to us, including Iran, North Korea, and Cuba. He envisions building a nuclear arsenal to deal with the Yankee imperialists. He is building his military forces and spending large sums of money to buy Russian weapons and aircraft.

North Korea:

North Korea has recently obtained nuclear status and has launched its first intercontinental ballistic missile. Its nuclear and missile programs are progressing rapidly. It now has the technology to be a serious threat to the United States. Although it does not have a large stockpile of missiles or nuclear weapons, with current technology, it would take only one nuclear explosion high above the United States to produce a powerful EMP (electromagnetic pulse) that would disable a large area of the country. Most importantly, it would disable our defenses. It would deliver a devastating blow to our security and throw our standard of living back to the preindustrial era. With two or three well-placed nuclear high atmosphere explosions, the entire country could be devastated.

Iran:

Iran is currently working feverishly to develop similar technology. Iran's president, Mahmoud Ahmadinejad, is a radical Islamic fundamentalist who believes that the Mahdi is coming soon and that it is his responsibility to prepare the world for his coming by destroying Western civilization and converting the entire world population to Islam. His plan is either to convert or to kill all the infidels with whatever means are available. He has vowed to destroy both Israel and the United States.

Mr. Ahmadinejad assumed the title of President of Iran in August, 2005. He has defied the United Nations by continuing to develop Iran's nuclear capabilities. Estimates by experts of how soon Iran will have nuclear weapons vary, but most believe it will be very soon. Experts predict that once he has nuclear weapons at his disposal, he will not hesitate to use them against U.S. cities and military installations.

Pakistan:

Pakistan has a population of 173 million, an army larger than that of the United States, and an arsenal of a hundred nuclear weapons. Much of the terrain is mountainous and difficult to control. Al-Qaeda is deeply entrenched in Pakistan and is a constant threat to the government. Extremist violence is spreading across the country. Terrorism experts fear that the country is at risk of falling under Al-Qaeda control. They warn that the U.S.-friendly Pakistani government could collapse within months and the country may disintegrate into Islamic warlord-run fiefdoms supporting Al-Qaida and other terrorist groups.

An Al-Qaeda-controlled Pakistan is one of terrorism experts' worst nightmares. With control over Pakistan's nuclear arsenal, the threat from Al-Qaeda to not only the United States and its allies, but to India, the Persian Gulf, and Central Asia would rise significantly.

In short, we are reaching a perfect stormlike situation with worldwide overpopulation, economic decline, ideological differences, and environmental pollution. Technological advances in weaponry give dangerous power to countries once considered insignificant in terms of military power. The ability to incapacitate or even destroy entire cities or nations is, or soon will be, at their disposal. Those who are prepared for such an event are the most likely to survive.

D. Precursors to Natural Disasters:

Global Warming:

There is a great deal of controversy regarding global warming. The areas in dispute include 1) whether it is real or perceived because of faulty measurements, 2) whether greenhouse gases really are the cause of increased global average air temperatures 3) whether global warming will continue to increase or will it reverse to a period of cooling, and 4) what the probable consequences of global warming are.

Is this warming trend unprecedented and due to human activities, or is it a result of normal climatic variations? Proponents of the former theory argue that pollution by man is responsible for global warming and that if it continues unchecked, much of the earth will eventually become uninhabitable. They propose sweeping programs to cut greenhouse gases by radically decreasing industrial and automotive pollutants. They advocate strict governmental regulations, controls, and increased taxes—measures they believe would help implement their programs.

Proponents of the latter theory argue that climate changes constantly over the millennia, illustrating how the Great Lakes area was covered by glaciers ten thousand years ago. They argue that the sun's activity has a great deal of influence on the earth's climate and that any recent increase in global temperature could be solely due to a cyclical increase in solar output. They also argue that much has been made of the subject for political gain, that the consequences of global warming are greatly overstated, and that the subject has been used to frighten the public into accepting otherwise unwanted political ideology and increased taxes. They argue that the sun's activity has recently abated somewhat and that colder times might emerge in the near future.

When scientists cannot agree, it is nearly impossible for the average citizen (including legislators) to determine the definitive cause of global warming. The outcome is indeed a political football. The Cap and Trade Bill is being pushed by politicians and their Wall Street bedfellows, including retired Goldman Sachs executives who have managed to occupy the highest government positions (read, "The

Great American Bubble Machine," by Matt Taibbi, *Rolling Stone Magazine*, 1082-1083, July 9–23, 2009). Both groups have their own agendas; the politicians anticipate greater tax revenues to spend and greater power of government while the ex-Wall Street execs anticipate billions in revenue for their old alma maters and greater retirement pensions from fees generated by trading the pollution rights on Wall Street.

Taxing American citizens and manufacturing companies for carbon footprints is an illogical idea. It will only drive more jobs out of the country. The effect of such a tax on global warming is highly suspect. The only way this tax (Cap and Trade) would lower worldwide greenhouse gas pollution is if the tax is imposed by every country on every citizen and manufacturing facility in the world. But even then the effluence of chemicals with proven harmful effects would continue unabated.

The best course of action is to understand the controversy and make rational preparations. Whether natural or man-made, the consequences of global warming are predicted to include an increase in natural disasters, including drought, famine, floods, hurricanes, forest fires, and tornadoes. Only time will tell whether the prediction is right.

E. Natural Disasters:

Natural disasters caused by the forces of nature include famine, epidemics, earthquakes, floods, forest fires, hurricanes, landslides, tornadoes, tsunamis, volcanic eruptions, winter storms, and heat waves. Proponents of global warming believe natural disasters will increase as temperatures rise.

Disasters can be man-made, natural, or a combination of the two. Famine, epidemics, forest fires, and landslides fall in this dual-cause category.

Famine:

The word "famine" refers to a widespread shortage of food in a geographical area. A famine usually results in malnutrition and starvation in the affected population; this in turn leads to epidemics of communicable diseases that result in increasing mortality. Famines can occur anywhere, any time. Historically, famines have occurred because of agricultural problems such as drought, floods, crop diseases, or insect infestation. Wars, acts of terrorism, governmental blunders, and economic decline have also caused what are known as man-induced famines.

Hundreds of famines occurred in Europe during the Middle Ages. More recently, one of the worst famines occurred in North China in 1876–79, when the death toll was almost thirteen million. A famine occurred in India about the same time with five million deaths.

During the twentieth century, it is estimated that seventy million people died from famines worldwide; as many as thirty-six million died in China alone during the famine of 1958–61, which occurred as a result of the Maoist Great Leap Forward policies that led to a decrease in food production.

The Holodomor is the famine that took place in Soviet Ukraine during 1932–33. It has been attributed to failed agricultural policies of the Soviet Union; it caused the deaths of up to ten million people. Other famines of the twentieth century occurred in Ethiopia (1983–85), where the cause was a combination of war and drought, and in North Korea (1990s), where there were unprecedented floods, tidal waves, and drought.

Food distribution problems also result in prolonging famines. During the Nigerian civil war (1967–70), the Nigerian government created a famine in the secessionist state of Biafra by blockade of supply routes. This government action resulted in the death by starvation of up to one million people.

Let's examine one particular famine in more detail. The Irish potato famine during the midnineteenth century was caused by a fungus disease that turned potatoes into black, inedible mush. The fungus, *Phytophthora infestans*, causes a disease known as "late blight."

The population of Ireland in the mid-1800s was eight million. It was an agrarian society, and the Irish were among the poorest people in the Western World. Only about a quarter of the population was literate. The average life expectancy for men was only forty years. It was common for girls to marry at sixteen and boys at seventeen years of age.

At the time of the blight, the Irish countryside was largely owned by the Protestant English and Anglo-Irish. They were the recipients of the lands that been confiscated from the native Irish Catholics in the mid-1600s by Oliver Cromwell. The new Protestant landlords used local agents to manage their estates while living lavishly in London or in Europe. Farmland was rented to Catholic farmers indirectly through middlemen, who divided the larger areas into smaller ones.

Around the late 1500s, potatoes were introduced to Ireland. Farmers quickly discovered the vegetable thrived in their country's cool moist soil and required very little labor. The plants were prolific. An acre of fertilized potatoes could yield as many as twelve tons of the tubers, enough to feed a family of six for a year with leftovers to feed the farm animals. Potato became the main staple food for more than half the population.

Between 1800 and 1845, Ireland had experienced regional and short-lived famines with only modest loss of life. The failure of the potato crop in September, 1845, was disastrous. The crop failure became a national issue for the first time, affecting the entire country. The blight turned the leaves on the potato plants black and curled. The disease also attacked the potato itself, making it inedible. The spores of the fungus travelled rapidly in the wind, so that a single infected plant could spread the disease and infect thousands more plants in just a few days. The blight infection lasted for six years.

Economic policies also played a part in the potato famine. At the same time, other grains and foodstuffs, which were not affected by the fungus, were being exported from Ireland rather than used to feed the starving farmers and villagers because of the high prices received from other countries.

One million people died of starvation. Many more died from epidemics of typhus, tuberculosis, cholera, scurvy, and dysentery. Those who could, escaped to other countries like the United States, Canada, and Australia. By 1851, Ireland lost more than two million due to starvation, disease, and emigration.

Throughout human history, famines have been confined to regions or countries. But now there are signs that famines may soon be global. Agricultural lands across the world are decreasing due to urbanization, climate change, drought, and deforestation. An even more serious fact is that the production of oil, crucial to modern agriculture and transportation of foodstuffs around the world, is peaking and will start to decline while the world population continues to increase.

With the current state of the world economy, the depreciating value of the U.S. dollar, high unemployment, and recession/depression in many parts of the world, it is likely that food prices will rise dramatically, resulting in a global economic famine. Food will become so expensive that many people will not be able to afford it.

Social disintegration accompanies famine and economic decline as people resort to distressed sales of assets in order to buy food. When their assets are gone, many turn to crime.

Taking into account both the unpredictable causes of natural famines and the predictable causes of man-made ones, it seems prudent to have a food storage program. Given the current state of world and national affairs, including economic decline, looming social unrest, rumblings of civil disorder, and the proliferation of weapons of mass destruction, now is the time to begin accumulating adequate supplies if you want to survive in this dangerous world.

Epidemics:

As long as people lived in small isolated groups, epidemics were sporadic and limited. As people began clustering in cities, they shared water and food supplies, handled unwashed food, and were exposed to primitive communal garbage and toilet facilities. Epidemics have occurred since man began living in communities. There are many communicable diseases that can turn into epidemics. Among the most devastating are bubonic plague, influenza, smallpox, typhus, cholera, and tuberculosis.

Black Death (Bubonic Plague):

In AD 1347, the bubonic plague, also known as Black Death, swept through Europe, killing at least one-third of the population. During the Middle Ages, an average of eight to thirteen million people died annually worldwide. The mortality rate for those afflicted ranged from 30 to 75 percent. In this disease, victims develop fever with temperatures ranging from 101 to 105°F, headaches, nausea, vomiting muscle aches, and inflamed lymph nodes (called buboes), particularly in the groin, armpits, and neck. Buboes become swollen and tender and can be as large as an egg. They can break open and discharge foul-smelling pus.

There are two other less common forms of the plague: pneumonic and septicemic. The pneumonic form has a mortality rate as high as 95 percent; it can occur from inhaling infected droplets from infected patients and from spread of bubonic plague through the bloodstream to the lungs. The septicemia (blood stream) form was the least common form, but it had a mortality rate approaching 100 percent. For both the bubonic and septicemic forms of the plague, there is bleeding and multiple system failure followed by death occurring between three to seven days from the onset of illness.

Plague is caused by the bacterium Yersinia pestis. It is spread to humans by fleas that live on rodents, including rats, prairie dogs, squirrels, and mice. The disease is particularly likely to occur in areas of famine and social disintegration where garbage and sewage collect, allowing rats to breed extensively.

The Great Plague of England (1665):

London's population was decimated during the Great Plague of 1665. That year the city had experienced a very hot summer. Sanitation facilities were almost nonexistent. Garbage and human waste were thrown out into the streets. The black rat population exploded. The disease spread with frightening speed. Those who could, left London for the comparative safety of the countryside. The first victims of the plague lived in the slum areas of London. There the population density was high, and people could not avoid contact with either rats or infected humans. Families that had one member diagnosed with the plague were quarantined in their homes for forty days and nights. A red cross was painted on the door to warn others to stay away.

Searchers were hired to find dead bodies or plague victims who were yet to be found by the authorities. The collected bodies were then put on carts and transported to mass burial pits. The plague was at its peak in September 1665. The onset of colder weather slowed the spread of the disease. The worst had passed by the end of 1665, but the plague did not end until September 2, 1666, when the Great Fire of London devastated the filthy city areas where rats had thrived.

Today bubonic plague is still endemic in certain areas. Madagascar, Tanzania, Brazil, Peru, Burma, and Vietnam have experienced cases almost every year since the last pandemic in 1880. Prairie dogs in the southwestern United States still carry the disease, and an occasional case has been reported there in humans.

Influenza:

Hippocrates was the first to record an influenza pandemic in the year 412 BC. Since 1580, there have been thirty-one additional flu pandemics recorded.

Influenza is caused by a virus that is readily spread by aerosol droplets from the mouths and noses of infected persons. Symptoms include a sudden onset of fever, chills, headache, sore throat, cough, muscle aches, and sometimes a feeling of exhaustion. In the northern hemisphere, the disease usually peaks from December to March.

The mortality rate from the milder strains of influenza is usually less than 0.1 percent, but because the number of patients infected is so high, there is a significant number of deaths, usually due to pneumonia or other pulmonary complications. In a typical year in the United States, twenty thousand people die of bacterial pneumonia as a secondary infection to flu. During heavy epidemics caused by a more virulent strain of the virus, the mortality figures can rise significantly.

The 1918 Influenza Pandemic:

The influenza pandemic of 1918 caused an estimated twenty-five to fifty million deaths worldwide. The pandemic hit in three waves. The first wave began in March, 1918, and spread through the United States, Europe, and Asia over the next six months. Death rates during the first wave were not unusually high. The second wave, in the fall of 1918, spread across the globe and was highly fatal. The third wave occurred from February to April, 1919. Both the fall and winter waves had a much higher frequency of cases complicated by severe pneumonia. It has been postulated that the virus underwent spontaneous changes in its genetic makeup between the first and second waves, accounting for the increased severity of the disease in the succeeding waves. This influenza epidemic was different in the pattern of mortality in that in addition to the usual high mortality rate in infants, young children, and the elderly, there was an unusually high death rate among persons under thirty-five years of age.

The death toll due to the influenza pandemic in the United States was 650,000. At its height, mortality rates were 15.8 percent in Philadelphia, 14.8 percent in Baltimore, and 10.9 percent in Washington DC. The rest of the world did not fare well either. Mexico had 500,000 deaths, Russia lost 450,000; Italy, 375,000; and Britain, 228,000. There were 44,000 deaths in Canada. Millions died in the Asian subcontinent.

World War I had already claimed nine million lives in its four-year duration. But a far greater and more lethal killer was lurking in the shadows.

On the morning of March 11, 1918, Albert Mitchell, a company cook at Camp Funston (now Fort Riley), Kansas, reported to the infirmary complaining of a low-grade fever, sore throat, headache, and muscle aches. By noon that day, 107 soldiers exhibited similar symptoms. Within two days, a total of 522 people were sick. Some were close to death with severe pneumonia. Other military bases reported similar figures almost immediately. By April, French soldiers and civilians were infected. Within two weeks, the disease had spread to China and Japan. By May, it had spread to Africa and South America. This first wave was more deadly than the typical flu but was still far from what it was yet to be.

The second wave was more virulent. Deaths from pneumonia were unusually high. Both military and civilian populations were decimated. The flu was a major determinant on the outcome of some battles. Many soldiers were too sick to engage in the fighting.

The 1918 influenza epidemic turned out to be the most lethal in human history. An estimated one third of the world's population of 1.5 billion (about 500 million people) was infected. Total deaths were estimated at fifty million though the toll may have been as high as a hundred million.

The 2009 Influenza Epidemic:

Will the current 2009 outbreak of influenza duplicate the 1918 pandemic? Will the first spring wave be repeated in the fall and winter? If so, will the second and third waves, like those of the 1918 pandemic, have significantly higher mortality rates?

With our current limited knowledge of the 1918 virus, predictions are only educated guesses. However, it has been estimated that even with antiviral drugs, vaccines, and antibiotics, which were unavailable in 1918, an influenza pandemic today, equal in pathogenicity to the 1918 virus, may kill more than a hundred million people worldwide.

The current status and recommendations regarding the 2009 H1N1 influenza pandemic can be followed at the Web site www.ready.gov through the link to the disease website.

Earthquakes:

The death toll from earthquakes is massive and staggers the imagination. The earthquake death toll in Eastern Sichuan, China, in 2008 was over 87,000. In 2005, an earthquake in Pakistan claimed 886,000 lives. Two hundred and twenty thousand people died as a result of the Sumatran earthquake of 2004.

The Richter Scale was developed in 1935 by Charles F. Richter. The scale determines the magnitude of earthquakes by measuring the seismic waves (vibrations) and comparing them on a logarithmic scale. The magnitude of an earthquake is expressed in whole numbers and decimal fractions. Each whole number increase in magnitude represents a tenfold increase in measured amplitude.

Earthquakes measuring 2.0 or less on the Richter Scale are usually called micro-earthquakes. They are not usually felt by people and are generally recorded only on local seismographs. Earthquakes measuring 4.5 or greater are strong enough to be recorded by seismographs around the world. Moderate earthquakes may have a score of 5.0, while a strong earthquake might have a score around 6.0 Great earthquakes have magnitudes of 8.0 or higher. One earthquake of this higher magnitude usually occurs somewhere in the world each year.

The Richter Scale does not assess damage. Earthquakes of equal magnitude may cause considerable loss of life and economic damage in densely populated areas.

In the United States, damage due to earthquakes has not been as massive as in other parts of the world. The San Francisco earthquake of 1906 caused the greatest number of deaths, estimated at three thousand. However, past experience is not a reliable predictor of future events, and earthquakes of varying magnitudes can occur at any time particularly along the Pacific rim stretching across the West Coast, Alaska, and Hawaii.

Floods:

Floods are natural or man-induced occurrences. Natural phenomena such as climate change, hurricanes, weather systems, and snowmelt can cause floods. Man-induced floods are those caused by failures of dams and levees and by inadequate drainage in urban areas.

Improved warning systems and evacuation planning have lowered fatalities, but economic losses have increased due to increased property development in areas prone to flooding.

In the United States, around 160 deaths due to flooding occur annually in a typical year, and more than half of all fatalities during floods are auto related. Floods annually cause billions of dollars in damages. Although certain areas are more prone to floods than others, they can occur at any place. Storms and hurricanes are the principal causes of floods in the eastern United States and the Gulf Coast. Rainstorms and snowmelt are the principle causes of floods in the western United States.

The costliest natural disaster in the history of the United States was the flooding precipitated by Hurricane Katrina in 2005. Katrina caused more than fifteen hundred deaths and over two hundred billion dollars in damages. The costliest river-related flood occurred in the Midwest in 1993, and caused twenty billion dollars in damages.

Floods can occur rapidly and with little warning. Rapid evacuation of an area is often necessary, leaving little time for gathering supplies, personal items, and mementos. Having an emergency backpack is particularly relevant if you live in a flood-prone area.

Volcanoes:

Currently active volcanoes have the potential of killing tens of millions of people and devastating entire countries. Two of the most active volcanoes in the world, Mt. Etna and Mt. Stromboli, both in Sicily, would affect most of Europe, particularly Italy and Greece, should massive eruptions occur. Unanticipated volcanic eruptions have taken an enormous toll on life and property throughout history.

A brief account of the most notable eruptions will put the potential risk in perspective.

Santorini:

The largest volcanic eruption in the past ten thousand years occurred about 1650 BC. Santorini, a small volcanic island in the Aegean Sea off the coast of Greece, erupted and then collapsed. The tidal wave that followed was over a hundred feet high and swept over the Mediterranean crescent. The death toll has been estimated to be in the millions. The Minoan civilization on the island of Crete was devastated, as were other civilizations in the area.

Mount Vesuvius:

Probably the most well known of the historic volcanic eruptions was that of Mount Vesuvius. The eruption occurred on August 25, AD 79. A 750°F cloud of hot gases swept down from the volcano and enveloped the town of Pompeii, a resort village, killing the residents and holiday visitors instantly with thermal shock. The town was then covered with lava. Clouds of ash rained down on the nearby town of Herculaneum, inhabited by wealthy Romans, burying it and its people under seventy-five feet of volcanic material. It has been estimated that at least ten thousand people died as a result of the Mount Vesuvius explosion.

Mount Tambora:

Indonesia's Mount Tambora erupted on April 5, 1815. It was also one of the largest eruptions in history. In Indonesia itself, 83,000 died. Sulphur-rich gases spewed to a height of twenty-eight miles and created major climate changes over a large area. The spring and summer of 1816 were extremely cold across Europe and North America. Snowfalls and frost occurred during the summer, killing most of the crops. Destruction of the corn crop resulted in farmers slaughtering their livestock. The resulting famine resulted in at least a hundred thousand people starving to death.

Krakatoa:

Krakatoa had a massive eruption on August 26, 1883. The total energy released by the Krakatoa explosions has been estimated at two hundred megatons, ten thousand times more powerful than the atomic explosion at Hiroshima. Two-thirds of the island was blasted fifty thousand feet into the atmosphere. The ensuing tidal wave was 120 feet high and swept over Indonesia, destroying thousands of villages and killing an estimated 36,000 people. The ash and rock particles blocked the sun for two days. The filter effect of the dust affected the world's weather for five years, causing the earth to cool. Following a period of dormancy, the volcano is once again active.

Mt. St. Helens:

The most recent volcanic eruption of major consequence was that of Mount Saint Helens in the State of Washington. The volcano had been dormant since the 1850s. The May 18, 1980, eruption killed fifty-seven people, destroyed 250 homes, and caused over a billion dollars in damage. A plume of volcanic ash rose to a height of eighty thousand feet and spread over eleven states. There was a massive landslide along with the eruption, which reduced the height of the mountain by 1,300 feet and destroyed everything in its path.

Mt. Kilauea:

A similar threat is posed by Mt. Kilauea on the island of Hawaii. There have been thirty-four eruptions since 1952, and it has continued to be active since 1983. Kilauea ranks among the world's most active volcanoes. There is concern that, in a way similar to what happened at Mt. St. Helens, part of the mountain may break off from the rest of the island and slide into the ocean, causing a tsunami of monumental size.

Forest fires:

The greatest forest fire in U.S. history occurred in 1871. The town of Peshtigo, situated on the Peshtigo River in northeastern Wisconsin, was totally destroyed in that conflagration. It was a booming town with a population of 1,700. The town's main industry was manufacturing wood products and boasted the largest woodenware factory in the country. The swirling fire that engulfed the town destroyed every building and claimed at least 800 lives in and around the town. The total number killed in the extensive fire has been estimated at 1,300. More than one million acres were burned across Wisconsin and Michigan.

In terms of total area burned, the largest fire in North America occurred in 1825. The fire raged from Maine through New Brunswick, Canada. Started inadvertently by a group of loggers who ignited a fire in a drought area, the wildfire burned over three million acres and claimed more than 160 lives.

The terms forest fire, firestorm, and wildfire have been used almost synonymously. However, a forest fire is defined as an uncontrolled fire occurring in vegetation more than six feet in height. An extensive forest fire may spread through the topmost branches of the trees before involving the undergrowth or the forest floor. As a result, a firestorm may develop, with violent tornadolike whirls that develop as hot air from the burning undergrowth rises. Such a fire is uncontrollable and subsides only upon the total consumption of the forest and underbrush. A wildfire is simply defined as a sweeping and destructive conflagration especially in a wilderness or rural area.

Forest fires can be naturally occurring or man made. Naturally occurring forest fires can be started by lightning strikes, especially in drought-afflicted areas. Man-made fires are those caused by careless campfire maintenance or deliberate acts of arson. Forest fires spread rapidly and may require quick evacuation of an area.

Hurricanes:

Hurricanes are intensely powerful cyclones that originate at sea in tropical waters. The term hurricane is reserved for those storms occurring over the North Atlantic Ocean. In the western North Pacific Ocean they are called typhoons, while around Australia and the Indian Ocean they are called tropical cyclones. These storms are characterized by circular wind patterns, where violent winds spiral around the eye of the storm. They can be hundreds of miles wide.

The deadliest hurricane to hit the United States was the one that devastated Galveston, Texas, in 1900. Casualty figures have been estimated to range from 8,000 to as many as 12,000. It was classified as a Category four hurricane. The second most deadly hurricane was the one that hit Lake Okeechobee, Florida, in 1928 and killed 2,500 people.

Hurricanes cause damage with their intense winds (with speeds greater than 74 mph), high waves, strong currents, flooding, and storm surges (domes of ocean water as high as 20 feet extending along the shore for up to 100 miles). Hurricanes also cause damage with associated tornadoes, landslides, and coastal erosion.

The severity of a hurricane is measured by the Saffir-Simpson Scale which ranks storms on a scale of 1–5, with 5 being the most severe. Storms are ranked by their wind speed, and type of damage expected from the storm.

Category 1: Minimal, 74–95 mph: Some damage is expected, with most of it limited to shrubbery, unanchored houses, and items. Some minor flooding will cause pier damage.

Category 2: Moderate, 96–110 mph: Considerable damage can be expected to shrubbery, and some trees may be blown down; there will be damage to mobile homes, signs, roofs, windows, and doors. Small craft may be torn from moorings, and marinas will probably flood. Low-lying areas and shoreline residences should be evacuated.

Category 3: Extensive, 111–130 mph: Large trees and most signs may be blown down; there may be structural damage to small buildings; mobile homes will be destroyed. Serious flooding will occur at the coast, with severe damage to shoreline structures and flooding up to eight miles (13 km) inland at elevations of five feet (1.5 m) or less.

Category 4: Extreme, 131–155 mph: Expect trees, signs, and traffic lights to be blown down, and extensive damage to be done to roofs, windows, and doors. Mobile homes will be completely destroyed. Beaches will be eroded, and there will be flooding as far as six miles (9.5 km) inland for anything under ten feet (3 m) above sea level. Anyone staying within five hundred yards (457 m) of shore will be evacuated, as will all single-story residences within two miles (4 km) of shore.

Category 5: Catastrophic, 156+ mph: Trees, signs, traffic lights will be blown down. There will be extensive damage to buildings and major damage to lower floors of structures less than fifteen feet (4.5 m) above sea level within five hundred yards (457 m) of shore. Massive evacuation of residential areas 5–10 miles (8–16 km) from shore will be required.

Weather forecasters issue a *hurricane watch* when a tropical storm intensifies, and it appears likely that a hurricane will develop within twenty-four to thirty-six hours. A *hurricane warning* is issued when hurricane conditions are expected for the area within twenty-four hours. Evacuation orders are frequently issued for areas with hurricane warnings.

Landslides:

By definition, a landslide is the movement of a mass of rock, debris, or earth down a slope. It commonly occurs in conjunction with other natural disasters such as severe storms, floods, earthquakes, and volcanic activity. They can also occur because of changes in groundwater and disturbance or change of a slope by man-made construction activities.

Landslides can occur and cause damage in all fifty states and U.S. territories. They cause between twenty-five and fifty deaths annually and one to two billion dollars in damages.

The worldwide death toll per year due to landslides is much higher. Catastrophic landslides, usually associated with other natural events, have claimed many thousands of lives. Most people are killed by rock fall, debris flows, or volcanic landslides. In 1998, Hurricane Mitch hit Central America and caused the deaths of approximately ten thousand people through a combination of flooding and landslides. The Haiyuan landslides associated with an earthquake in China in 1920 killed a hundred thousand people. The USGS lists twenty-six catastrophic landslides worldwide during the twentieth century.

The world's biggest recorded landslide occurred during the 1980 eruption of Mount St. Helens, a volcano in the Cascade Mountain Range in the state of Washington. The volume of material was 2.8 km³.

Tornadoes:

A tornado is a rotating, funnel-shaped cloud that extends from a thunderstorm to the ground with winds that can reach 300 mph. In the United States, tornadoes are most likely to form in an area known as tornado alley, which extends from the Rocky Mountains to the Appalachians, and from Iowa and Nebraska to the Gulf of Mexico. Tornadoes can, however, form in any state and cause considerable damage and loss of life.

Tornadoes are nature's most violent storms. They may strike quickly with little or no warning, leaving behind a path of death and destruction. The average tornado has a speed of about 30 mph, and usually moves southwest to northeast. However, tornadoes may move in any direction.

From 1950 to 2004, the average number of tornadoes per year was 910. During the same period, the states which witnessed the highest number were Texas, Oklahoma, Kansas, Florida, and Nebraska. The states with the lowest number of tornadoes were, from least to most, Alaska, Rhode Island, Hawaii, Vermont, and Oregon. The month with the most tornadoes reported was May, 2003; it recorded 543 tornadoes. The most deaths reported in one year was 519 in 1953. The year with the fewest deaths was 1986, with fifteen deaths reported. Overall, May and June are the months with the greatest number of tornadoes. However, the greatest number of deaths is reported in April.

The deadliest tornado in the United States occurred in 18 March, 1925, in the tristate area of Missouri, Illinois, and Indiana. The death count was 689 and another 2,000 were injured. Property damage was estimated at $16.5 million. In 1840, a tornado struck Natchez, Mississippi, killing 317 and injuring over 1,000 residents. Another deadly tornado hit St. Louis, Missouri on May 27, 1896, killing 255 people. At least 82 people died as a result of tornadoes in 2008.

The standard rating system for tornadoes is the Fujita Tornado Intensity Scale. This system classifies tornadoes according to the damage they inflict. The scale ranges from F-0 to F-5.

F0=light damage: 40–72 mph winds

F1=moderate damage: 73–112 mph winds

F2=significant damage: 113–157 mph winds

F3=severe damage: 158–206 mph winds

F4=devastating damage: 207–260 mph winds

F5=incredible damage: 261–318 mph winds

F-1 is the most frequent category, accounting for slightly less than half of all tornadoes. F-1 tornadoes are capable of turning over mobile homes and automobiles as well as tearing roofs from houses and uprooting trees. F-5 tornadoes are capable of lifting houses and carrying them significant distances, such as *The Wizard of Oz* tornado.

Weather forecasts usually issue warnings when tornado conditions are present. They will issue either a *tornado watch* or a *tornado warning*. A tornado watch means that weather conditions are such that a tornado is possible and that you should remain alert for approaching storms, watch the sky, and stay tuned to your radio or TV for further weather alerts. A tornado warning indicates that a tornado has been sighted or seen on the weather radar. If outdoors, you should take shelter immediately.

Winter Storms/Blizzards:

The Great White Hurricane that occurred between 11 and 14 March, 1888, was the most severe winter storm to ever hit the Northeast. Up to fifty inches of snow fell in parts of New Jersey, New York, Massachusetts, and Connecticut. Winds of over 45 mph created snowdrifts as high as fifty feet. People were trapped in their houses for up to a week. Trains stopped running. Property loss from fires was estimated at $25 million. Many people, trapped in their homes, died of starvation.

A more recent storm in the Northeast was the Great Blizzard of 1978. Twenty to forty inches of snow fell over a thirty-three-hour period amidst winds of 65 mph. Motorists were stranded on highways and several people died while trapped in their automobiles. Snowdrifts up to fifteen-feet high trapped people in their homes and offices.

Winter storms occur regularly across the country, causing great inconvenience, property damage, and death. These storms can produce power outages for several days or more, resulting in loss of heat, refrigeration, and the use of other electric appliances. Roads may become impassible for days. People caught in winter storms without proper shelter and/or clothing rapidly become hypothermic and die.

The four most common types of winter storms are blizzards, ice storms, lake effects, and northeasters.

Blizzards

Blizzards are characterized by low temperatures (usually below 20°F), winds that are at least 35 mph, and falling and/or blowing snow that reduces visibility to 1/4 mile or less for at least 3 hours.

A severe blizzard has temperatures near or below 10°F with winds exceeding 45 mph. Visibility is reduced to near zero by falling and blowing snow.

Ice Storms

Ice storms are characterized by freezing rain that accumulates to at least 1/4 thickness. Freezing rain results when rain droplets land on freezing or subfreezing temperatures on the ground or objects near the surface. Ice accumulates on roads, tree limbs, and power lines, creating hazardous driving conditions, breaking tree limbs, and downing electric and telephone lines.

Lake Effect

Lake effect snows occur when a mass of cold air moves over a large body of warm water such as the Great Lakes. Clouds form over the water and eventually deposit heavy snows as they move inland. Winds accompanying the cold air masses generally blow from the west or the northwest, causing the lake effect snow to fall on the east or southeast sides of the lakes.

Northeaster

Northeasters are so named because they are caused by strong northeasterly winds blowing in over coastal areas from the ocean ahead of the storm. They are usually ferocious storms and may form off the East Coast of the Atlantic Ocean or in the Gulf of Mexico. Northeasters produce high winds, heavy snow, rain, and oversized waves that cause beach erosion and property damage.

Winter precipitation is of four types: rain, freezing rain, snow, and sleet.

Rain

All four types of precipitation start out as ice or snow crystals in clouds. If these ice or snow crystals pass through a layer of warmer air as they descend toward the ground, they melt into rain. As long as the air remains above freezing for the remainder of the descent, rain will fall at ground level.

Freezing Rain

Freezing rain results when rain droplets land on the ground or on objects near the surface which have freezing or subfreezing temperatures.

Snow

Snow crystals in the clouds fall to the ground as snow when temperatures are below freezing in all or most of the atmosphere from the cloud to the ground. Snow flakes are six-sided ice crystals. Ten inches of snow will melt into one inch of water.

Sleet

Sleet is precipitation falling as small pellets of ice. Ice pellets occur when snowflakes melt into raindrops as they first pass warmer air. The raindrops then refreeze into ice particles as they fall through subfreezing air near but above the surface of the earth. The difference between freezing rain and sleet is that freezing rain occurs when raindrops fall into subfreezing air which is so shallow the raindrops do not have time to refreeze into ice until they make contact with the ground.

Heat Waves:

Heat waves can affect anyone. Infants, young children, elderly, and people with medical problems (particularly heart and kidney diseases) are more susceptible to increased temperatures and humidity than are healthy adults. In the forty-year period from 1936 through 1975, nearly twenty thousand people died in the United States from the effects of heat and solar radiation.

A heat wave is a prolonged period of excessive heat and humidity. The National Weather Service devised the heat index to accurately quantify the danger from a particular heat wave; the displayed number in degrees reflects the effect of air temperature and relative humidity acting together. The heat index was devised for shady and light windy conditions. Full sunshine and strong, hot, dry winds can raise the index as much as fifteen degrees.

For example, with prolonged exposure and/or physical exercise, a heat index of 80–90 degrees may cause fatigue. A heat index of 90–105 degrees may cause sunstroke, heat cramps, or heat exhaustion. At 105–130 degrees, sunstroke, heat cramps, or heat exhaustion is likely, and heatstroke is possible. At 130 or higher, heatstroke and sunstroke are highly likely.

Heat disorder symptoms include sunburn, heat cramps, heat exhaustion, and heatstroke (sunstroke). Heat-wave safety precautions include limiting physical activity, wearing lightweight clothing, eating lightly, drinking plenty of fluids, and spending as much time in air-conditioned areas as possible. Moreover, one needs to stay out of direct sunlight and avoid fluids with alcohol or caffeine.

Heat-related conditions and symptoms:

Sunburn: Sunburn is characterized by skin redness and pain. In severe cases, blisters form, and there may be swelling of the skin in addition to fever and headaches. Treatment consists of applying sunburn ointment/cream with lanolin or aloe, and a topical anesthetic such as lidocaine. If blisters break, apply dry, sterile dressings. Severe cases of sunburn should be seen by a physician.

Heat cramps: Heat cramps are muscular pains and spasms caused by heavy exertion in hot weather. They are the least severe of heat-related problems. However, heat cramps are an early sign that the body is having trouble adjusting to the heat. Treatment includes moving the patient to a cooler place where he/she can rest in a comfortable position, giving fluids (cool water every fifteen minutes), and lightly stretching the affected muscles. The victim should avoid alcohol or caffeine during this time.

Heat exhaustion: People who develop heat exhaustion are dehydrated. They have heavy sweating, cool, moist, pale, or flushed skin, headache, nausea, and/or vomiting, dizziness, and exhaustion. Body temperature is usually near normal. People develop heat exhaustion while physically exerting themselves in an area with a high heat index. If untreated, the condition may progress to heatstroke. Treatment of heat exhaustion consists of removing the person from the heat and into a cooler place. Remove or loosen clothing and apply cool compresses to the skin. Give a half glass of cool water to drink slowly every fifteen minutes. Avoid liquids that contain alcohol or caffeine. Let the victim rest in a comfortable position, and observe carefully for changes in his or her condition.

Heatstroke/Sunstroke: Heatstroke is a life-threatening emergency. People who develop heatstroke have body temperatures as high as 105°F. They have hot, red skin that may feel wet or dry (wet if the person was sweating from exercise or heavy work), a rapid, weak pulse, and rapid, shallow breathing. They may or may not be conscious. The heatstroke patient's temperature-control mechanism stops functioning. The body temperature can rise high enough to cause brain damage, and death may result if the body is not cooled quickly. Immediate medical treatment is required in such cases. Call 911, or transport the patient to the nearest emergency room. If hospital care is not immediately available, move the person to a cooler place, and quickly decrease the body temperature by immersion in a cool bath or by wrapping wet sheets around the body and fanning. If the patient is conscious, give cool water to drink slowly. Make arrangements for hospital transport and care as soon as possible.

Tsunamis:

Tsunamis are water waves caused by sudden vertical movement of a large area of the seafloor during an undersea earthquake. The waves travel rapidly (sometimes over 400 mph) to distant shores, and as they reach shallow water, produce giant waves as high as eighty feet or more. These waves have caused tremendous loss of life and property.

Tsunamis have hit the West Coast of the United States in the recent past. The 1946 Aleutian tsunami produced waves of twelve to sixteen feet at Half Moon Bay, Muir Beach, Arena Cove, and Santa Cruz, California. In 1960, a tsunami originating in Chile produced waves of twelve feet at Crescent City, California. Interestingly, the 1906 San Francisco earthquake produced local tsunami waves of only several inches. The largest known locally generated tsunami on the West Coast of the United States was the one produced by the 1927 Point Arguello, California, earthquake. That tsunami produced waves of about seven feet in the nearby coastal area.

Alaska

The earliest recorded tsunami in Alaska was in 1788. The 1964 earthquake produced a tsunami that killed 107 people and caused damage worth $80 million in the area. The same tsunami travelled to California and produced waves over twenty-feet high at Crescent City, causing $7.5 million damage and eleven deaths. Waves ranging from ten to sixteen feet occurred along the coastlines of California, Oregon, and Washington.

Hawaii

About 90 percent of all recorded tsunamis have occurred in the Pacific Ocean. Over one hundred tsunamis have been recorded in the Hawaiian Islands since 1819. Sixteen of these have caused significant damage. The worst locally generated tsunamis occurred in 1869 and 1975 on the southeastern coast of the big island of Hawaii. The thirty-five-feet high tsunami at Hilo in 1960 was generated by an 8.3 magnitude earthquake in Chile. There were sixty-one recorded deaths and damages worth millions of dollars in the area. Much of the coastal portion of Hilo was destroyed in that tsunami.

In short, geographical areas differ in their susceptibility to the various types of natural disasters. However, they can occur with or without warning any place at any time. If this account of disasters has not stimulated you to start a preparedness program, take the vulnerability index test to see how unprepared you really are.

PART 2. VULNERABILITY INDEX

The first step in preparing and planning a preparedness program is to discover how vulnerable you are to death or injury during a disaster. I have developed both a questionnaire and a scale to determine your vulnerability. The questions are weighted in value so that the most important areas have a higher influence on your total score. You can, from your answers, determine which areas you need to consider as priority areas for improvement.

The Vulnerability Test:

Your vulnerability index test score will be calculated from the answers you provide to the following questions. The index calculation weighs the answers to the various categories according to their importance for your survival under disaster conditions. The questions are grouped by category. Each category will have one to eight questions. The categories are emergency backpack, food storage, water supply, shelter, medical preparedness, and miscellaneous.

After you have answered each question, we assign a number to your answer. When all the numbers have been assigned, we will add the numbers, and the total will be your vulnerability index score. The higher your score, the more vulnerable you are to disaster situations and the more likely you are to die, be seriously injured, suffer extreme stress, and possibly develop post-traumatic stress syndrome or just be very uncomfortable. The lower your score, the less vulnerable you are and your chances of survival are better.

Circle the appropriate answers to the following twenty questions which start on the next page. Read the discussion following each question, and note the values (in parentheses) assigned to your answers.

1. Do you have an emergency backpack?

- (15) No

- (0) Yes

This question rates highly in significance because if you have to leave your premises in a hurry, you won't have time to gather any food, water, medications, and medical supplies to take with you, and you could find yourself very uncomfortable after even a few hours. It is worth fifteen points or 15 percent of the total vulnerability index score.

Emergency situations that may require evacuation of an area include chemical, biological, and nuclear attacks or accidents, as well as floods, forest fires, and other natural disasters. Evacuation orders may come suddenly and require immediate action.

I cannot stress enough the importance of every person having an emergency backpack. It is a three-day lifeline. It should be fully packed and ready to go at a moment's notice. You can keep it (or a second backpack) in the trunk of your car in case you are at work or on the road at the time a disaster strikes. If there is an emergency evacuation order, you will have to leave your home and all your survival supplies behind.

Your emergency backpack should contain as many supplies as you can comfortably carry. When possible, keep items in airtight plastic bags or boxes.

1. Food: A three-day supply. Food should be lightweight and nonperishable. MREs (meals ready to eat), freeze-dried pouches, dehydrated foods, radiated pouches, and canned goods are best. Replace items at least once a year.

2. Water: The heaviest item in your backpack. Water weighs about eight pounds per gallon. Many will not be able to carry as much as required. For home storage and use, the ideal amount is two gallons per day per person or six gallons for a three-day supply. That amount would add forty-eight pounds to your backpack. Even one gallon per day would add twenty-four pounds. Using water conservatively, an adult can get by with one-half gallon (~2

liters) per day for drinking, and under extreme water-shortage conditions, an adult can survive on one quart (~one liter) per day for several days. Using these figures, the *minimum* amount of water in your backpack should be three to six one-liter bottles weighing between six and twelve pounds (one liter weighs ~two pounds). The more you are able to carry, the better. Six 1 liter bottles will weigh about twelve pounds. Carry up to twelve one-liter bottles (or more) if you can tolerate the weight and have room. Exchange the water bottles in your backpack every six months.

3. Medications: Prescription and nonprescription. Take *at least* a one-month supply of all prescription and nonprescription medications. Keep them in an airtight and watertight plastic bag.

4. First-aid kit: Should include at least the following items: sterile gloves, sterile dressings, antibacterial soap, antibiotic ointment, cortisone cream, burn ointment, adhesive bandages, eyewash solution, a thermometer, any special health-monitoring devices for blood pressure or blood glucose measurement, and a minor surgical kit. Commercially available compact first-aid kits are good for backpacks. Add several surgical face masks.

5. Clothing: One complete change of clothing. Thermal underwear in cold climate. Don't forget rain gear (lightweight plastic poncho and lightweight plastic sheeting).

6. Toiletry supplies: Toothbrush, toothpaste, soap, bathroom tissue, shaving supplies, comb, and feminine hygiene products.

7. Radio: Compact multiband, battery-operated, crank generator, or solar powered.

8. Flashlight: With extra batteries, or crank generator powered.

9. Whistle.

10. Moist towelettes: Individually packaged.

11. Matches: In a waterproof container.

12. Plastic bags: Ziplock bags are useful to keep item such as matches and clothes dry.

13. Swiss Army Knife: A useful item for your backpack with its variety of small tools, including a can opener, screwdriver, and knife blades.

14. Cash: At least several hundred dollars in smaller bills.

15. Important documents: Personal identification (passport, driver's license, etc.).

16. Weapon: Necessary to ward off those who would steal your goods (a likely situation in a time of civil disorder).

17. Small utensil: A cup and small kettle to hold/heat water.

Don't forget your cell phone!

Individually adapt the backpacks for your infant (formula, diapers, etc.), small children, and people with disabilities to meet their special needs.

Remember, your backpack will supply you with basic needs for three days, whether you are home, on an evacuation route (on which there may be long delays), in the country with a stalled car, or in an evacuation center.

2. Have you started a food storage program?

- (15) No

- (10) Yes, I have enough food stored to last from two to four weeks

- (5) Yes, I have enough food stored to last from one to three months

- (3) Yes, I have enough food stored to last from three to six months

- (1) Yes, I have enough food stored to last more than six months

Having a food storage program is one of the primary elements of preparedness. Whatever the disaster, natural or man-made, you will most likely be cut off from outside food sources. When a disaster is anticipated, grocery store shelves empty quickly due to panic buying. Once disaster strikes, transportation of food supplies will be severely impaired or may stop completely. Supermarkets will close. You will have to depend solely on what you already have in the pantry. You should have at least a month's supply of nonperishable food readily available at all times. The longer your supply will last, the better.

A food storage program is critical to your survival. A person can live without food for a week or even longer, but he/she will be very uncomfortable with the hunger pangs and the resulting weakness.

There are a number of decisions to make regarding your food storage program. You have to calculate a budget: How much do you want, or can you afford, to spend on your emergency food reserve program? What kind of food do you want to eat during an emergency? Can you count on refrigeration during an emergency? Probably not.

You must decide what items to buy. Will you be happy eating Campbell's tomato soup, or would you rather have lobster bisque? Will you have a stove available to cook food? Don't count on frozen foods. There is a good chance that you will lose electricity and the food will spoil quickly. In general, buy food supplies that you would usually eat. Buy food items that do not require cooking unless you have a small one or two-burner propane camp stove, a few canisters of propane fuel, and basic cooking utensils. Be sure you have a can opener available!

Packaged cereals, prepared pasta dishes, canned meat (Spam), fruits, and vegetables are good choices. They have a shelf life of one to two years and are easily stored. When buying canned goods, always look at the expiration date on the can or bottle. Some stores carry products that have already been stored for a few months. You do not want canned goods that expire in just a few months. You can rotate the foods before their expiration date. If you buy in bulk amounts, go to discount stores like Sam's Club or Costco to save money. Keep in mind how much storage space you have available. Look for unused spaces such as under the bed or a closet floor.

MREs are meals sealed in specially designed pouches and then cooked. After cooking, they remain sealed until eaten. They have a shelf life of up to ten years when stored at temperatures between 60 and 70°F.

You may prefer to buy freeze-dried foods. They have a shelf life of thirty-five years or more when they are sealed in cans. The pouches (enough for one meal for one or two people) have a shorter shelf life. The #10 cans usually hold enough food for twenty or so meals, lasting one person approximately one week. If you have not eaten freeze-dried food, know that it is easy to reconstitute. Just add water (preferably hot) and wait several minutes. There is a great variety of freeze-dried foods to choose from, and some are tastier than others. They are easily stored, and their long shelf life is a big advantage.

Dehydrated foods are generally less expensive than freeze-dried foods, but the shelf life is shorter, usually only one to two years. Some are available in sealed cans, others in plastic pouches. A variety of dehydrated foods is available, including fruits, vegetables, scrambled eggs, milk, and jerky. Now radiated meats are also available, including bacon, tuna, and other meat products. Check the labels for expiration dates.

Make sure you have a variety of foods in your program. Don't forget cooking essentials such as cooking oil or shortening, as well as nutritional supplements, including vitamins.

If you are just starting your food storage program, buy what you can afford now. If you cannot afford a six-month or one-year supply, buy what you can and add to it later. The more you can buy now, the better. Prices will probably go up with inflation. Your food needs will vary with the type of disaster. At least start with a three-day supply for your backpack. That will get you through a short-term disaster, or at least buy time until outside help arrives. Remember to eat the foods in your refrigerator and freezer first when disaster strikes. They will spoil quickly if and when the electricity goes off.

There are a number of suppliers on the Internet, which have assembled long-term freeze-dried/dehydrated food combination bundles for sale. Most of these companies also have other preparedness items available as well.

If you live in the country, growing your own vegetables and fruits can provide a large amount of food for longer-term storage programs, but it requires considerable work and frustration battling insects, marauding raccoons, groundhogs, and other pests. Raising animals is another option, but it requires space and a lot of care. On the positive side, knowing that you have a steady supply of fresh eggs, chicken, milk, beef, or other animal products can be very reassuring.

What did our ancestors do to survive when there was no electricity to run freezers or refrigerators? They preserved foods when they were in season by canning them. The old-fashioned method of canning involves placing the fruit or vegetable in a Mason jar, heating it in a pressure cooker to high temperatures, then placing a flat metal seal on the jar while it is still hot. The heat and pressure kills the bacteria and mold spores, and as the jar cools, the air inside contracts and produces a vacuum inside the jar. As long as the seal is intact, the food remains edible for many months or even years. Details of the process can be found on the Internet.

There is another canning process called dry pack canning which is used for low-moisture food products such as grains (rice, wheat), powdered milk, powdered eggs, pasta, dried beans, dried fruit, and other foods with little water content. The process involves placing the product in a can (#10 cans are frequently used) to just below the top and either adding an oxygen-absorbent packet to the can or filling the can with an inert gas such as nitrogen or carbon dioxide. The can is immediately sealed using a machine designed for that purpose. Depending on the food product, it is preserved for two to seven years safely. This method does not work with high moisture, high fat, or oily food products. Details of this process and the machinery required can also be found on the Internet.

Another source of food during emergencies is hunting. This subject is covered in another section.

If you live near a clean body of water and it is safe to be outside, fishing can supplement your diet with fresh protein. Bait is usually readily available (worms, crickets, other larger bugs) under rocks or in soil. If you do not know how to fish, it can be an enjoyable pastime and an invaluable skill to learn. Knowledge of edible plants can also provide you with in-season delicacies such as nuts, wild berries, mushrooms, and even weeds, such as dandelions. There are many plants or parts of common wild-growing plants that are edible, even tasty, but you must be familiar with them. Many plants are poisonous, so you must know how to recognize them as well.

3. Have you started a water storage program?

- (8) No
- (4) Yes, but only enough to last up to two weeks
- (3) Yes, but only enough to last from two weeks to four weeks
- (1) Yes, but only enough to last one to three months
- (0) Yes enough to last more than three months, or I have a shallow well with a hand pump.

Water is a critical factor in being prepared for disasters. I weighted the water storage program at eight points and your water source (next question on the V.I. test) at seven points on the vulnerability index, giving your total water program a total of fifteen points. Without water, we quickly become dehydrated and develop electrolyte imbalance as well as other physiological problems. Death may occur after just a few days. Water is vital to your survival.

Water is easily bought and stored in plastic containers of various sizes, ranging from six to eight-ounce bottles to five-gallon containers. Overall, either the one-gallon or the 2 ½-gallon size is the best choice for most home storage programs, although a supply of smaller containers for portability is desirable, especially for your backpack. It is cheaper to buy water in larger vessels, but when you get to the five-gallon size, they become quite heavy and bulky. Water in one-gallon plastic jugs generally costs slightly over a dollar a gallon. Glass bottles are the best for maintaining water quality, but they are heavier and bulkier than plastic. Use only food-grade plastic bottles. Make sure your water containers are tightly sealed and stored in a cool area away from toxic chemicals. Rotate your stored water at least annually.

Each person needs one gallon a day for basic survival, two to three quarts for drinking/cooking (more if the weather is hot and/or you are exercising heavily), and the rest for basic hygiene and other purposes. If you want to survive with a few niceties, count on two gallons a day per person.

If you are just starting a water storage program, I suggest buying a three-day supply in smaller containers for your emergency backpack and at least fourteen gallons per person for your longer-term storage program now; you should add to it periodically as you raise your preparedness level. Remember, with your food storage program, you have to find a place to store the containers. They can be heavy. One gallon of water weighs about eight pounds.

4. What is the source of your water?

- (3) Municipal water supply

- (2) Well without backup generator

- (1) Well with backup generator

- (0) Adequate supply of bottled or other storable water

Dependence on a municipal water supply carries a high score because it is vulnerable to shutdowns from power failures and loss of infrastructure (broken pipes), and it is a potential target for terrorists who can inject poisons into the supply.

Having your own well lowers the risk considerably, but deeper wells require electricity to run a pump. Having a backup generator for the well lowers the risk considerably. Having an alternate source of water (shallow well with hand pump, large water storage tank, or cistern) also lowers the risk in this category. However, everyone should have a supply of bottled water in reserve.

If you are caught in a situation with no water, there are sources in your home that can be used in an emergency. Your water heater may contain up to sixty gallons of water. There will also be water in the pipes that can be drained and saved if you know where the drain valve is located. Swimming pools can be a source of large amounts of water. During rainstorms, rainwater can be collected from gutter runoffs. Water from these sources should be treated with water purification tablets or by boiling before drinking. Avoid drinking water that is from a source contaminated with potentially toxic chemicals.

Water sources outside the home include nearby ponds, streams, rivers, and lakes. Depending on your location, these sources may be contaminated with agricultural and chemical runoff, toxic chemicals from other sources, and/or bacterial contamination from septic discharge. Bacterial contamination can be treated by boiling the water for fifteen minutes and/or adding chlorine bleach (two to four drops per quart of water) or tincture of iodine (three to six drops per quart of water) and waiting forty-five minutes for the bacteria to be killed. Commercially available water-purification tablets may also be used for this purpose. Chemical contamination is harder to remove. Charcoal and other filters are commercially available for this purpose and are desirable to have among your emergency supplies.

5. Where do you live?

- (4) City 1: large city (over one million population)
- (3.5) City 2: smaller city (less than one million population)
- (3) Suburb
- (2.5) Small town
- (2) Rural 1: less than five acres
- (1.5) Rural 2: 5–20 acres
- (1) Rural 3: farm (over twenty acres)

The location of your residence is a significant factor in the vulnerability index, accounting for 4 percent of the total points and 27 percent of the fifteen residence-related points (questions 5–12).

Where you are at the time of a disaster is of paramount importance. Most likely, it will be where you live or where you work. We stress the home environment because most people spend the majority of their time at their residence. If your work location is significantly different from your residence, you must calculate a vulnerability index for your work location as well.

There are seven residence location categories from which to choose. Living in a large city carries the highest risk because these areas are the most likely locations for terrorists, particularly biological, chemical, nuclear, or suicide bombers. Cities are also the most likely locations for civil disorder and riots. In addition, in cities, you are totally dependent on public utilities, such as electricity, water, and waste disposal. Impaired transportation in high population density areas will result in shortages of food and other supplies.

On the other hand, if you live on a farm, you are relatively self-sufficient. Food is readily available. With a generator, there will be electricity available to run your well, and you will have your own septic system. You will be far away from the terrorist attacks, civil disorder, riots, and looting that will be mostly confined to cities. However, in the event that rural areas become targets of raiding city dwellers looking for food and supplies, it is important to have the means to defend yourself.

If your job confines you to a city location, a second home or other secure place in a rural area should be considered. If this is impractical, make arrangements to stay with a relative or friend who lives in a more secure area. If there is adequate warning of a disaster, or you are highly suspicious that one is about to occur, you will have time to get away from the higher-risk area. Keep in mind that both places should be stocked with emergency supplies (food, water, etc.), your vehicle should always have at least one-half tank of gas, and your emergency backpack should be readily available.

6. What is your residence type?

- (3) High-rise building (condo or apartment)
- (2.5) Manufactured home
- (2) Low-rise building (condo or apartment)
- (1.5) House without basement
- (1) House with basement

There are five residence categories to choose from, ranging in descending order from the most vulnerable to the least. Vulnerability in this category depends on the type of disaster, so ranking here is more subjective. I have made an apartment or condo in a high-rise building the most vulnerable because of the potential for collapse (consider the World Trade Center buildings) and the inability to get in or out in case of power failure (which disables the elevator) or fire. There is the additional problem of getting supplies in when there is elevator failure.

Because of the structural weakness of manufactured homes, I ranked them the second most dangerous place to be. These structures are the most likely to be damaged or destroyed in a tornado or hurricane, or from a blast of any kind.

A basement adds a good measure of protection in terms of offering both a secure area and protection from tornadoes and high winds. On the other hand, a basement is a liability during floods. If you have a burglar/fire alarm system in your home, deduct one point from the above rating score for your home type.

7. Do you have a secure area at home (where you can safely hide your goods and yourself)?

- (3) No
- (2) Yes, hidden room in house
- (2) Yes, hidden room in basement
- (1) Yes, hidden above ground shelter on premises
- (1) Yes, underground room/bunker
- (0) Yes, underground room/bunker, self-contained with recirculating air system

A secure area is critical in times of disaster. It should be structurally sound to withstand natural disasters. It should provide concealment of yourself and your family from marauding burglars and looters during civil disorder. It can also serve as a storage area for your survival supplies. The safe area can be a concealed area of your basement, a hidden room in your house, or an above or underground shelter. The amount of protection you derive from your secure area is dependent upon how much time, effort, and expense you put into it. For secure, long-term use in times of NBC warfare or other prolonged disasters, many thousands or hundreds of thousands of dollars will be required to build a secure above or underground shelter complete with an independent power supply, heating and ventilating system, water and plumbing systems, and appropriate internal environment monitors. The proper and safe design and construction of this type of long-term shelter requires hiring experts with knowledge and expertise in these areas.

A safe room provides protection and security for you and your family during severe weather conditions, attempted break-ins, civil disorder, and other disaster situations. Ideally, it should be fireproof, blast proof, and should have a concealed entrance. Its location and entrance should be unknown to all except those who use it. For longer-term occupancy, it should have electric power, adequate lighting, an air-circulating system, a heating system, a dehumidifying system, and bathroom facilities. There should be ample room for storage of food, water, a tool kit, and other supplies, including a weapon and ammunition for defense.

A safe room can be as simple as a barricaded area in your basement or as complex as an underground chamber complete with the amenities described above. There are several prebuilt safe rooms available. They can be transported to your site and placed according to your specifications. They are, of course, rather expensive. Small, bare-bones storm shelters are available for as little as $4,000–5,000. More elaborate systems designed to withstand NBC warfare and other long-term calamities can be priced at over a million dollars.

8. What type of heating fuel do you use for your home?

- (4) Electric heat
- (4) Forced air-natural gas
- (4) Forced air propane
- (4) Forced air-fuel oil
- (3) Steam heat
- (0) Wood stove

Your ability to heat or cool your home can be a critical factor in your survival. Without heat in winter, a family can freeze to death. In warm climates, lack of air-conditioning can make life uncomfortable. The elderly and debilitated are particularly vulnerable to temperature extremes.

The scoring on this question is based on the likelihood of the loss of power for your heating system. In times of disaster, a heating system that depends on electricity is the most vulnerable. This includes not only electric heat, but also forced air systems, which depend on a fan to circulate the air. These systems are vulnerable to grid shutdown or downed power lines, conditions that are to be expected during both man-made and natural disasters. Homes heated with natural gas are also vulnerable to shutoff during a disaster. Steam heat is dependent on the availability of the type of fuel used for the furnace. A wood stove is the least vulnerable; it provides radiant heat, and the fuel is only as far away as the nearest tree. If you have a wood stove, it is prudent to have an adequate supply of firewood already available.

9. Do you have a backup heat source for your home?

- (1) No
- (0.5) Yes, pellet stove
- (0.5) Yes, kerosene heater
- (0.5) Yes, propane heater
- (0.5) Yes, coal stove
- (0) Yes, wood stove or fireplace

A backup heating system can be life-saving if you live in a cold climate. Any type of backup heating is desirable, but some are more dependable than others. Pellet stoves require electricity to run the worm drive that carries the pellets into the burning chamber. Kerosene, propane, and coal, all require periodical replenishment of the fuel supply, as does wood, but wood is most likely to be available when delivery trucks are not running.

A wood stove is the ideal backup heating source, since fuel is readily available in most locations. A fireplace offers some protection, but it is less efficient than a wood stove.

10. Do you have a generator or alternate power source (solar, hydroelectric, wind)?

- (1) No

- (0.5) Yes, gasoline or diesel powered

- (0) Yes, solar or wind powered

If a major disaster occurs, you can bet the electricity will be off indefinitely. You will have no lights, no furnace, no water unless you have a hand-pump well, and no refrigeration. A reliable generator can be a lifesaver and can at least provide you with the usual comforts of your home. Gasoline, propane, and diesel generators are dependent on renewing the fuel supply, and eventually you will run out. However, with a large storage tank and conservation of fuel (use the generator only several hours a day), a finite supply can be made to last a long time.

Solar- or wind-powered systems with storage batteries can provide electricity indefinitely but are expensive. Information on home solar and wind systems is available on the Internet, but learn the details before you buy. Some systems are useless during a power outage. They do not have battery storage capacity and work only if you already have electricity. Their selling point is that you lower your electricity bill. You have to use that type of system a long time before it will pay for itself. The worst feature of this type of system is that you will not have electricity when you need it the most. Be sure you buy a complete system with battery storage capacity.

11. Do you have one or more oil lamps or other alternate lighting source available?

- (1) No

- (0) Yes

Having an alternate source of light is part of a good preparedness program. If the electricity is off, you will not be able to function efficiently in the dark. The two most common alternate light sources are oil lanterns and battery-operated flashlights or lanterns. Be sure you have an adequate supply of oil or batteries.

If you use oil lamps, remember that they can be a fire hazard. Keep them out of reach of children and pets.

12. What kind of system do you have for sewer and waste disposal?

- (1) Municipal sewer system and garbage collection
- (0) Septic tank/home incinerator

Although not usually a life-threatening problem, lack of waste disposal services can be very unpleasant. Unflushed toilets get pretty rank after several days. Using disposable plastic bags to line the toilet can be a great solution to this problem. Remove them after several uses, secure them tightly, and place them in an out-of-the way place for storage until they can be removed from the premises.

Piled up garbage can smell pretty bad, too. It can attract rats and other vermin, as well as become a breeding ground for disease-causing bacteria. If there is no garbage pickup, the best solutions are incineration or burial. If neither is possible, place garbage in plastic bags as far away as possible from your living area.

13. Do you have a serious chronic disease or disability?

- (5) Yes

- (0) No

This and question 14 are really about medical preparedness. Each is worth five points. Being in the best possible health and physical condition is paramount for survival in an emergency. If you have health problems or special medical needs, you must be prepared to meet these needs at all times.

If you have not had a physical examination for more than a year, you should have one, complete with laboratory tests to determine whether or not you have any developing health problems (i.e., diabetes, kidney disease, high blood pressure, etc.). You should have monitoring and testing supplies (blood glucose meter/strips, sphygmomanometer to check blood pressure, etc.) available if you have any of these conditions.

You should also have your eyes checked and your glasses prescription updated, if needed. If you wear glasses, it is a good idea to have an extra pair.

Having a toothache during times of emergency can be both irritating and distracting, not to mention dangerous if it is due to an abscess. You should be sure you have regular dental checkups in order to keep the risk of dental complications at a minimum.

Disaster conditions frequently cause stress-related symptoms, including headaches, gastrointestinal problems (heartburn, acid reflux, diarrhea, irritable bowel syndrome, constipation), anxiety, depression, panic attacks, and elevated blood pressure. Stressful situations can also cause abnormal heart rhythms, difficulty sleeping, difficulty concentrating, irritability, muscle tension and soreness, and shortness of breath. These symptoms are a serious and widespread problem during disaster situations. High stress and anxiety levels can predispose vulnerable patients to heart attacks, perforated stomach and duodenal ulcers, and other potentially fatal conditions and complications. Just having confidence that you are properly prepared for disaster conditions can alleviate many of these stress-related symptoms.

The medical care system may disintegrate with a major disaster. Despite considerable planning and preparation, a major event will overwhelm medical facilities and health care personnel. Depending on the nature of the disaster, hospitals may be filled to capacity and close their doors, or they may be damaged or destroyed beyond function. Doctors, nurses, and other health care personnel may be killed or disabled. You must be prepared to survive without hospitals and doctors.

You will need a medical supply kit and the knowledge to recognize and treat common health problems. There are medical supply kits available on the Internet. A Google search will reveal a variety of them in various price ranges and complexity of contents. Study the quality of each kit and only buy a kit with items you can trust to work when you need them. You can also design your own kit and buy items individually.

There are several short courses available to increase your knowledge and skills in basic health care. The American Red Cross offers a certification course in BLS (Basic Life Support), which will train you in resuscitation techniques. Having these skills help you to remain calm during a medical emergency. There are also courses, manuals, textbooks, and DVDs available in first-aid and other medically related topics pertinent to survival in disaster situations. A Google search will lead you to a variety of videos on the subject. An Amazon search will lead you to a variety of books on these subjects.

There are, of course, limitations to what you can do under dire circumstances. With proper knowledge, equipment, and medications, you will be able to survive certain health emergencies, but others, such as acute life-threatening surgical conditions, severe trauma, heart attacks, and strokes that require intensive hospital and physician care, will frequently result in death. You must be mentally prepared for this possibility.

14. Do you have a supply of prescription medications (if needed) and other medications for at least three months?

- (5) No

- (2) Yes, enough to last three to six months

- (0) Yes, enough to last more than six months

If you take prescription medications on a regular basis (for high blood pressure, diabetes, etc.) you should have at least a three-month supply. A one-year supply is much better. Drug stores may be closed and doctors unavailable.

You will need basic medications to treat common ailments. Medications you should have available fall into two categories: prescription and nonprescription. Prescription drugs must be prescribed by a physician or other qualified health care professional. Make a list of 1) those drugs you or your family members take on a regular basis (blood pressure medications, diabetes medications, etc.) and 2) those medications you should have available to treat conditions which may occur under disaster conditions, such as broad-spectrum antibiotics for infections. Then discuss your plans for an emergency medical kit with your doctor. He should guide you and provide you with the necessary prescriptions.

Remember, hospitals and doctors are prime targets during terrorist attacks. Also if there are mass casualties and infrastructure damage, you probably won't be able to get near a hospital, a doctor, or a drug store.

For over-the-counter medications, a trip to the drug store will provide you with everything you need. You should also have a supply of nonprescription medications for any intermittent health problems you may experience, such as headaches, arthritis flare-ups, or stomach and intestinal problems. A well-equipped medical kit should include at least the following: aspirin, ibuprofen, Benadryl, Immodium, a laxative, at least one type of antacid (Tums, Zantac, etc.), an eyewash, solution and a bottle of hand sanitizer. Since air contamination is a problem during certain disasters, a supply of surgical or other high-quality masks should be available for each member of the family.

You should also have a well-equipped first-aid kit among your medical preparedness supplies. A basic kit should have at minimum the following: bandages (Band-Aids, sterile 4´4s, gauze); surgical adhesive tape; a thermometer; antiseptics (alcohol or Betadine swabs, hydrogen peroxide); a minor surgical kit with basic surgical instruments (scissors, tweezers, sutures, needle holder, forceps, scalpel); antibacterial soap; antibiotic ointment; sterile surgical gloves; adhesive dressings; and non-sterile disposable gloves.

15. Do you have a supply of personal hygiene products for at least one month?

- (5) No

- (3) Yes, enough to last from one to two months

- (1) Yes, more than two months supply

If disaster strikes, will you have personal hygiene products available? Will you have bathroom tissue available? Soap? Will you be able to brush your teeth? Take at least a sponge bath? (That's where the extra gallon of water per day comes in handy. A normal shower uses about two gallons per minute, so in an emergency a sponge bath is a great water saver.)

You should have *at least* a one-month supply of personal hygiene products. Three months is better, and six months to one year is better yet.

This category, although not generally dealing with life-threatening supplies, certainly can make a difference in your comfort level. Envision having to do without toothpaste or dental floss for a month or two. Or worse yet, bathroom tissue. Or feminine hygiene products. Being without soap or deodorant can get pretty rank after a while, also.

Here is a list of personal hygiene supplies you should consider for your storage program:

1. Toothbrush, toothpaste, and mouthwash

2. Dental floss or tape

3. Soap, washcloths, and towels

4. Bathroom tissue and Kleenex and Q-tips

5. Clothing items

6. Feminine supplies

7. Bed linens, pillow, and blankets for your cot/mattress

8. Shaving cream, razor blades, comb, deodorant

Start with at least a month's supply of these products to get to Level 1B on the preparedness scale (see next section). Then increase the quantity of your supplies periodically to rise to the next level.

16. Do you have cash available in case your bank closes?

- (5) Only enough to last less than two weeks

- (3) Yes, enough for two weeks to one month.

- (2) Yes, enough for one to three months

- (1) Yes, enough for more than three months

During times of disaster, don't expect the banks or stores to be open. Don't expect to use your credit card, either. You should have enough cash available to pay your bills (if conditions permit), buy groceries, gasoline, and other expenses for at least six months. This can amount to a sizable stash of cash, and it should be locked in a fireproof heavy safe or other secure location where it is safe from fire and burglars.

A month's supply of cash is better than no cash, but a six-month supply is better. You don't know how long the banks (or stores) will be closed. Your best source of buying, selling, or trading supplies may be friends and neighbors. You may be able to buy or trade a lot more using gold or silver coins than with cash as the dollar depreciates.

Don't keep your emergency cash reserves in your bank account or safe deposit box. Banks may close for a number of reasons, including economic collapse, civil disorder, terrorist attacks, and natural disasters.

17. Do you have emergency transportation in case you need to evacuate the area?

- (5) No

- (1) Yes (car, truck, SUV, motor home)

Certain disasters (biological, nuclear, chemical attacks, floods, forest fires, etc.) may require evacuation from an area. Have an advance plan regarding where to go. Staying with relatives or good friends living outside the disaster area is the best choice.

In case you have to evacuate an area, having your own vehicle with a full tank of gas will make your exit more efficient and more comfortable than waiting for and riding a bus. You will be able to take more of your stuff with you as well. For evacuation purposes, the larger the vehicle the better. A motor home is ideal because you can survive in the vehicle for many days with all the amenities. You have your own bathroom, stove, refrigerator and living quarters. Always make sure that your emergency vehicle's gas tank is *at least* half full.

18. Do you have communications equipment?

- (5) No
- (3) Cell phone and/or landline telephone
- (2) CB radio
- (2) Ham radio
- (2) Battery-powered or other alternate powered portable radio
- (0) Battery-powered or other alternate powered portable radio + one of the above

Emergency communications are essential during a disaster. You can receive up to the minute information about the disaster as well as communicate with your family members and friends, learning of their whereabouts and condition. You can make plans to meet at a certain location.

Landline and cellular telephones are excellent communication devices, but they are also the most vulnerable during a disaster, as the telephone wires and cell phone towers are subject to damage. CB and ham radios require electric power, but they are often powered by a motor vehicle battery. For incoming information only, a battery-operated or crank generator radio works well. The multiband radios are the best to have under these conditions. Local stations may be off the air, and short wave will inform you about the rest of the world.

You should have a preplanned communications program for your family members and friends. Telephones may be inoperable, and so a planned place to meet when disaster strikes or as soon as possible is essential. For local emergencies and for evacuation-requiring disasters, an out-of-town contact may provide the ability for all to keep in touch.

Depending upon the nature of the disaster, one of the first decisions you will have to make is whether to stay in the area or evacuate. You should have preplanned scenarios for both situations. If you need to evacuate immediately, each member of your family should have his emergency backpack ready to go.

19. Do you have readily available clothing for hostile environments (insulated underwear, rain gear, overcoat to match the climate)?

- (5) No
- (0) Yes

A disaster may force you to live outdoors for an extended period of time. You should have appropriate clothing available to survive in a hostile climate environment (cold, snow, rain, wind, etc.).

If a disaster strikes in winter and you lose your heating system, or if you are forced outside, you will rapidly become hypothermic and die in a few hours. It is essential to have the proper clothing to protect you under these circumstances. Items to consider are as follows:

1. Insulated underwear

2. Coats, hats and shoes/boots, and gloves suitable for local weather

3. Raincoat or poncho

4. Standard clothing items: extra shirts, pants, socks, and regular underwear.

20. Have you had a basic training course in the use of firearms?

- (5) No, and I do not own or have access to a gun.

- (4) Yes, but I do not own or have access to a gun.

- (2) No, but I own or have access to a gun.

- (0) Yes, and it included gun training, and I own or have access to a gun.

The lower scores for owning or having access to a gun assume that you also have ammunition for that gun.

Having the means to defend yourself from terrorists, predators, and others who would do you harm and/or steal your survival goods is critical to your survival during terrorist attacks, civil disorder with riots, looting, and those looking to rape and pillage during times of crisis. A gun can also provide you with the means to procure meat by shooting birds, rabbits, or other animals.

Many people will become desperate under extreme stress, especially if they have not planned ahead and are not prepared for disasters. If people know that you have supplies and they do not, they will turn to you for help. This can present a real dilemma for you. If you have enough extra to provide for them, be charitable and give it to them. However, if there is only enough for you and your family to survive, and giving some to others jeopardizes your survival, you may be forced to turn them away. Desperation may cause threats and the use of force to take supplies from you, in which case you will have to defend yourself and your family.

Others may not ask nicely, but just demand your supplies with the threat of physical harm to you and your family members if you do not comply. Suddenly you are forced into a survival of the fittest mode, and if you are smaller and weaker than your aggressor, you lose. If you have a gun, the great equalizer, you win. Hopefully just the threat of the gun will force the aggressor away, but if not, you may, under these desperate circumstances, have to shoot those who would rob you of your means of survival.

A gun is a basic survival tool. It not only wards off aggressors, robbers, and rapists, but it also provides you with food in times of emergency. If you are in the country, there are usually birds, rabbits, raccoons, and squirrels available to hunt. In the city, under starvation conditions, there will be stray pets that will become fair game. Eating them beats starving to death.

Anyone contemplating buying a gun should learn to use it safely and have the proper discipline to use it wisely. The National Rifle Association sponsors gun use and hunter safety courses throughout the country through sporting good stores, sportsman's clubs, and through private certified instructors. If you have not taken one or more of these courses, I highly recommend it. They also teach the legal ramifications of using firearms.

If you have no experience or training in hunting, here is some basic information. Different types of firearms are required for different types of game. For the novice, a .22 caliber rifle is a good start. It is best used for small game, such as squirrels, raccoons, opposums, and rabbits (especially if they are standing still!). Yes, you can eat these critters if you are hungry enough. Remember, a dead animal is useless unless you know how to skin it, clean it (remove the internal organs, etc.), and cook it.

The .22 caliber ammunition is the least noisy, which can be an advantage in many situations. There are three sizes of .22 ammunitions: shorts, longs, and long rifles. Long rifles are the noisiest but the most powerful, while shorts are the quietest but least powerful. If you don't want to attract attention, use the shorts.

The best type of gun for shooting birds is a shotgun. They come in four common sizes (from smallest to largest): 410, 20, 16, and the 12 gauges. The shotgun can also be used for shooting small game, but you may have to spit out the pellets in the meat occasionally. Shotguns are more noisy than .22s and they have a greater kick when you fire them. On the other hand, you are less likely to miss the animal with a shotgun.

A larger caliber rifle, such as a .30-.30, is better for larger animals such as deer, and is a better weapon for protection from predators.

If you find hunting offensive or disgusting, just remember that the meat you buy in the grocery store was once a live animal. You just didn't have to kill it, skin it, and cut it up yourself. But somebody else did. Under survival conditions, you can do it yourself.

You can also eat things you wouldn't normally eat. Most animals are edible, including horses, rats, raccoons, dogs, cats, and mice, disgusting as that may seem.

Calculate your Vulnerability Index Test Score

Now that you have finished scoring each of the twenty questions on your test, calculate your position on the vulnerability index by adding all your points on the test. The sum of all your question points is your vulnerability index score.

Interpreting your Vulnerability Index Test Score

The maximum possible score is one hundred points. A person with this score has the highest risk of death or serious injury in a disaster situation. The lowest possible score on the vulnerability index is 5.5. A person with this score has achieved maximum preparedness and is the least likely to die or be seriously injured in a disaster situation, including NBC Warfare.

Even the maximum amount of planning and preparedness is not a foolproof guarantee that you will survive a disaster situation. Certain disasters can be so overwhelming that no one in the area will survive. For this reason we structured the vulnerability index so that no one can have a perfect score of 0, which could be interpreted as total protection and safety from all forms of disaster situations.

The purpose of determining your vulnerability index score is to show you where to stand in the risk spectrum and to show you areas where you can improve (lower) your score significantly. Some areas may be too costly or change your lifestyle too drastically, while other areas may be very affordable with little change in lifestyle. Each individual must determine how seriously he feels the risks are, and to what extent he is willing to decrease his bank account and change his lifestyle.

Compare your score on the vulnerability index with the chart below.

Your Score	Vulnerability
80–100	Very high
60–80	High
40–60	Moderate
20–40	Low
5.5–20	Lowest

Now review your answers to the twenty questions and refer to the chart on the next page. Then look for areas which you can rapidly improve (lower) your score. Try to lower your vulnerability by improving the areas with the most influence on the score first. For example, if you do not already have an emergency backpack, you can lower your score by fifteen points by creating one.

The relative importance of each question is explained in the following chart:

RANK	CATEGORY	POINTS	WT. FACTOR
1	EMERGENCY BACKPACK	15	0.15
2	FOOD STORAGE	15	0.15
3	WATER STORAGE	8	0.08
4	WATER SOURCE	7	0.07
5	RESIDENCE LOCATION	4	0.04
6	RESIDENCE TYPE	3	0.03
7	SECURE AREA	3	0.03
8	HEAT FUEL	1	0.01
9	BACKUP HEAT	1	0.01
10	GENERATOR/ALT PWR	1	0.01
11	LIGHT SOURCE (OIL LAMP)	1	0.01
12	SEWER SOURCE	1	0.01
13	HEALTH STATUS	5	0.05
14	MEDICATIONS	5	0.05
15	PERSONAL HYGIENE	5	0.05
16	CASH	5	0.05
17	VEHICLE	5	0.05
18	COMMUNICATIONS	5	0.05
19	CLOTHING	5	0.05
20	FIREARMS TRAINING	5	0.05
	TOTAL	100	1.00

Now that we have calculated and reviewed your vulnerability index, let's explore another way to determine how prepared you are for disaster situations. For this, I have developed a Preparedness Level Scale.

PREPAREDNESS LEVEL SCALE

LEVEL 1: ABILITY TO SURVIVE FOR 1–4 WEEKS

* Category A) Basic Survival

* Category B) Survival with Comfort

* Category C) Able to Survive Severe Disasters Including NBC Warfare/Terrorism

LEVEL 2: ABILITY TO SURVIVE FOR 1–3 MONTHS

* Category A) Basic Survival

* Category B) Survival with Comfort

* Category C) Able to Survive Severe Disasters Including NBC Warfare/Terrorism

LEVEL 3: ABILITY TO SURVIVE FOR 3–6 MONTHS

* Category A) Basic Survival

* Category B) Survival with Comfort

* Category C) Able to Survive Severe Disasters Including NBC Warfare/Terrorism

LEVEL 4: ABILITY TO SURVIVE FOR 6–12 MONTHS

* Category A) Basic Survival

* Category B) Survival with Comfort

* Category C) Able to Survive Severe Disasters Including NBC Warfare/Terrorism

LEVEL 5: ABILITY TO SURVIVE FOR MORE THAN ONE YEAR

* Category A) Basic Survival

* Category B) Survival with Comfort

*   Category C) Able to Survive Severe Disasters Including NBC Warfare/Terrorism

The scale measures two factors: level of preparedness and time. An individual may have more than one rating. For example, if a person is prepared to survive comfortably for three weeks and prepared for basic survival for two months, the ratings would be Levels 1B and 2A.

The next step is to determine your current preparedness level. Every person should be prepared at least to the minimum of Level 1A. Obviously, the higher your level, the more likely you are to survive, and survive comfortably in a disaster. Once you have determined your current level, you should choose the level where you would like to be and then obtain the appropriate supplies to get there.

Keep in mind that preparedness involves both time and financial commitment. The higher the level of preparedness, the greater will be the expense. Basic survival at Level 1 may involve several hundred dollars. Reaching Levels 4 and 5 may involve many thousands of dollars. Attaining Category C at any level is the most difficult, most complex, and most expensive.

Preparing for an Uncertain Future

Now it is time to determine to what extent you want to increase your preparedness. Each person must determine in his own mind how likely it is that each disaster scenario will occur. Will there be a major conflict between the Globalists and the Nationalists? Or will they be able to come to some agreement whereby both sides can live with certain compromises? How likely is it that the Islamic fundamentalists will again attack us, perhaps on a much larger scale?

How likely is it that the North Koreans or Iran will attack us with one or two nuclear bombs in the atmosphere, reducing our standard of living and our defenses to preindustrial revolution levels? How likely is it that a tornado or hurricane will strike? Will global warming/climate change increase the risk of severe weather patterns? Will there be a total economic collapse accompanied by bank closures and civil disorder?

How much time and money are you willing to spend? Your answer depends on how seriously you take the threats and how much you can afford to spend. Some people will decide to make only minimum preparations, thinking that Level 1A is enough. Recall that Level 1A is able to survive for one to four weeks at the basic survival model.

Others with a very high level of concern and with unlimited resources may decide to buy a farm in a rural area, install an underground shelter, and store a year's supply of food, water, medication, and other supplies for themselves, their children, grandchildren, or people they care about the most.

One last note about supplies—whether you are buying food, containers, medical supplies, mechanical items, clothing, or other preparedness items, always buy the highest quality product you can afford. The last thing you need in a disaster situation is spoiled food, faulty medical products, or nonfunctioning equipment (i.e., your generator).

It is important to discuss preparedness with all members of the family. Make sure everyone knows where supplies are, when to go to your safe area, and what to do when the family is separated. *Make plans* and have mock drills.

It is my fervent hope that by reading this book you have 1) come to understand the risks we face, 2) developed a sense of your vulnerability, and 3) been inspired to take the appropriate measures to protect yourself and your family.

Preparedness is a complex undertaking. There are numerous resources for information and supplies on the Internet and I urge you to take advantage of them.

Don't wait until it is too late. Don't let it be another thing you waited too long to do. We live in a dangerous world, and the time to prepare for it, to protect ourselves, and our loved ones is now.

www.ingramcontent.com/pod-product-compliance
Lightning Source LLC
Chambersburg PA
CBHW030023290326
41934CB00005B/457